EARTH SCIENCE TODAY

EARTH SCIENCE TODAY

By
G. Singh
Reader
Dept. of Chemistry
C.C.S. University
Meerut (U.P.)
(India)

DISCOVERY PUBLISHING HOUSE PVT. LTD.
NEW DELHI-110 002

First Published – 2009

Reprinted – 2025

ISBN: 978-81-8356-438-0

Earth Science Today

Published by:

DISCOVERY PUBLISHING HOUSE PVT. LTD.
4383/4B, Ansari Road, Darya Ganj
New Delhi-110 002 (India)
Phone: +91-11-23279245; 23253475; 43596065
Mobile: +91 9811179893 / +91 9871656464
E-mail: discoverybooksindia@gmail.com
orderdphbooks@gmail.com
namitwasan9@gmail.com
web: www.discoverypublishinggroup.com

Printed at:
Infinity Imaging Systems

Contents

Preface

Earth Science is the study of earth and space. In other words, it is a term for the science, related to planet Earth. For the study of Earth Science, there are two approaches – reductionist and holistic. In earth science, four spheres are generally recognised, Lithosphere, hydrosphere, atmosphere and biosphere, which respectively correspond to rocks, water, air and life.

Beneath the earth's crest, lies the mantle, which is heated by the radioactive decay of heavy elements. This mantle is not quite solid and consists of magma, which is in the state of semi-perpetual convection. This convection process causes the lithospheric plates to move, albeit slowly and the resulting process is known as plate tectonics.

The features of earth can be explained by the actions of gradual processes, operating over long periods of time; for example, a mountain needs not be thought of having been created, in a moment, instead it may be seen as the result of continuous subduction, causing magma to rise and form continual volcanic arcs.

Realising the need of a worthwhile book on the subject; 'Earth Science Today' covers all aspects the subject. We have made this humble effort in order to fill the bill. Hopefully, the students, scholars and general readers would be benefited by this work. Positive feedback is awaited.

Evolution of Earth

Geologists' Views

The concept of isostasy, therefore, is fundamental to studies of the crust's major features— continents, ocean basins, and mountain ranges—and to understanding the response of crust to erosion, sedimentation, glaciation, and tectonic system.

Continental glaciers are a clear example of isostatic adjustment of the crust. The weight of an ice sheet several thousand metres thick disrupts the crustal balance and depresses the crust beneath. In both Antarctica and Greenland, the weight of the ice has depressed the central part of the land masses below sea level. Antarctica provides an excellent example of how continents maintain isostatic equilibrium in the face of changing conditions. Most of the southern continent is covered by a layer of ice as much as 4 km (13,000 feet) thick. The great weight of the ice has pushed much of the continental crust below sea level. If this ice were to melt, Antarctica would slowly rise from the water much the same way a ship rises while being unloaded. A similar isostatic adjustment occurred in Europe and North America during the last ice age, when continental glaciers existed there. Parts of both continents, such as Hudson Bay and the Baltic Sea, are still below sea level. Now that the ice is gone, however, the crust is rebounding at a rate of 5-10 metres per 1,000 years.

The construction of Bhakra dam is another example of isostatic adjustment because the added weight of water and sediment in the Gobind Sagar (reservoir) was sufficient to cause measurable subsidence. From the time of the construction of the dam in 1956 enormous quantity of water, plus an unknown quantity of sediment, accumulated in Gobind Sagar, behind the dam. Within a few years, this added weight caused the crust to subside around the lake.

In brief, the concept of isostasy depends upon the model of the earth's crust in which lighter, continental masses 'float' on a denser substratum. In the words of Steers (1961), the doctrine of isostasy states that whenever equilibrium exists on the earth's surface, equal mass must underlie equal surface areas.

The concept of isostasy has been conceived differently by the different geologists. The views of a few of them are discussed in the following paras.

Airy's Views: Sir George B. Airy (1801-92)—the Astronomer Royal in 1851—opined that the continents which are made of lighter material *sial* are floating over the substratum which is made of denser material *sima.* The suggestion was put forward to account for certain serious features concerning the gravitative attraction of the Himalayas. During the Trigonometrical and Geodetic Survey of India the difference of latitude between two stations, Kalianpur and Kaliana, was obtained both by direct triangulation and by astronomical methods. The difference in the latitude obtained by the two means amounted to 5.236. It is important to note that Kaliana, the northern of the two stations, is only about 100 km (60 miles) from the Himalayas. He opined that the lighter sial of the Himalayas is floating over the denser material of the sima lying underneath. He also made it vividly clear that the Himalayas were not just surface feature but that the light rocks of the mountain extended as roots deep down into the denser rocks below. Just as a large part of the boat remains immersed in water to keep it afloat, so are the Himalayas floating on the denser rocks below and the weight of the mountains is balanced by light materials penetrating into considerable depths.

In fact, the principle of isostasy is really nothing else than that of ordinary floatation, and the example of an iceberg floating in water may help stress the fundamental concept of the doctrine.

Ice floats in water in such a way that, approximately, for every one part above water level there are nine parts below. That is to say, the ratio of free board to draught is 1 to 9. On this analogy, if we assume the average density of the continental rocks to be 2.67 and that of the substratum as 3, as suggested by Joly, then nearly eight times the height of the rocks above the substratum should penetrate into the denser substratum. In other words, the density of the rocks must be relatively low down to considerable depths below these mountains.

The important point, however, is that they are compensated as a whole, and the individual peaks and valley are not separately compensated. To take a somewhat farfetched example, it would be quite wrong to assume that, if Mount Everest is about 30,000 feet high, and if the ratio assumed of draught to free board is correct, then there is a downward projection of lighter material beneath that mountain reaching to about 2,40,000 feet. By stating of mountain areas being compensated, it is implied that the mountains are regarded as a great plateau whose height is equal to that of the mean height of the mountain chain. The same reasoning must also apply to the continents. Joly estimates their mean heights and concludes that the continental material is about 31 km thick on an average.

Airy established his point with the help of an illustration of wooden blocks of different heights floating in water with each block moving vertically and independently of its neighbour. All blocks are of equal density but different heights, and they float, the longest blocks extending deepest into the mantle. At a certain depth, equal to or greater than the thickest crust, the pressure is everywhere constant and below it the mantle is in a state of hydrostatic equilibrium. The blocks emerge by amounts which are proportional to their heights, so that the block which rises highest above the water is also the block which goes deepest into the water. These blocks are said to be in a state of hydrostatic balance or equilibrium. Similarly, the land masses are composed of similar density of rocks, but they penetrate deep into the substratum in proportion to their height. Thus, Airy assumed that the continents and islands are resting, or rather floating, in denser mass, and that excess of matter above the upper surface of the substratum is balanced by deep projection of the lighter material into the substratum. In short, the sial masses are in hydrostatic equilibrium. The fundamental concept of Airy, i.e., the continental masses floating as lighter *(sial)* blocks in a heavier *(sima)* substratum, has been rejuvenated by Heiskanen's work, so that it is now probably true to say that most geologists favour Airy's explanation.

Pratt's Views: The data obtained from the trigonometrical and geodetic survey of Kalianpur and Kaliana in Indian plains were computed by J.H. Pratt. According to Pratt, the gravitational attraction of the Himalayas was less than the mass represented by these mountains because the mountains were made of much lighter materials, i.e., rocks of much lower density. He stipulated that the rocks of the mountain chains should have a lower density than those of the plateaus, the rocks of the plateaus lower than the plains and the plains lower than the rocks of the ocean floor. In his opinion, there was an inverse relationship between height and density—higher the column lesser the density, and smaller the column greater the density. Pratt visualised that there was a level above which these density changes were found and below that level the density was uniform. In other words, the density of a particular column was the same from top to bottom but the density of different columns were different.

Pratt's concept of isostasy was related to 'the law of compensation' and not to 'the law of floatation'. According to Pratt, different relief features are standing only because of the fact that their respective mass is equal along the line of compensation because of their varying densities. He, however, does not believe in the 'root formation'. Thus, the fundamental difference between Airy's and Pratt's views is that the former postulated a uniform density with varying thickness, while the latter a uniform depth with varying density.

Views of Hayford and Bowie: The views expressed by Hayford and Bowie about isostasy were in agreement with that of Pratt. According to them, there is a plane where there is a complete compensation of the crustal parts. Densities vary with elevations of columns of crustal parts above this plane of compensation.

Thus, the density of the mountains is less than the ocean floor. In other words, the earth's crust is composed of lighter material under the mountains than under the floor of the oceans. According to Hayford and Bowie, there is an inverse relationship between the height of columns of the earth's crust and their respective densities above the line of compensation. The plane or level of compensation is supposedly located at the depth of about 100 km. The columns having the rocks of lesser density stand higher than the columns having the rocks of higher density.

Joly's Views: The views advocated by Hayford and Bowie were criticised by Joly. His main objection was about the depth of compensation (100 km). In his opinion, the temperatures would be so high that all rocks would melt and become liquid at such a depth. According to Joly, there is a 16 km (10 miles) thick layer below the areas having similar low density. The areas of low density penetrate into this thick layer, while where the surface rocks are of higher density, the heavier materials below rise to higher levels. Joly, thus, conceived the level of compensation not as a straight line surface but as a zone of compensation. This view is much closer to the floatation theory of Airy. In the opinion of Joly, the earth's crust consists of lighter materials *(siat)* whose average density is 2.67. Below this is the *sima* whose density is 3. The sial is floating over the sima like an iceberg. Thus, these views are in conformity with the model of the earth's crust in which the lighter continental masses 'float' on a denser substratum.

Holmes' Views: The views of Arthur Holmes were also similar to that of Airy. Holmes was of the opinion that the higher relief features of the earth's surface are composed of lighter materials and their portions lie buried in the heavier and denser rocks below to keep them in a state of hydrostatic balance. He further opined that mountain ranges have roots largely composed of sialic rocks, going down to depths of 50 to 60 km. In coastal plains the thickness of the sial and other crustal rocks is only 30 km or less and beneath large parts of the deep ocean floor, the sial is either extremely thin or totally absent. According to Holmes, the lighter columns are standing because of the fact that there is lighter material below them for greater depth whereas there is lighter material below the smaller columns upto lesser depth.

Adjustment of Isostasy

A perfect isostatis adjustment is possible only theoretically. In fact, the earth is unstable and unresting as the endogenetic and exogenetic forces continuously disturb the isostatic adjustment. In fact, the endogenetic forces and the resultant tectonic events cause disturbances in the ideal condition of isostasy but the nature always tends towards the isostatic adjustment, if not at the local, at the regional level. For example, a newly formed mountain due to tectonic activities is subjected to severe denudation. Consequently, there is continuous lowering of the height of the mountain. On the other hand, eroded sediments are deposited in the oceanic areas, with the result there is continuous increase of weight of sediments on the sea floor. Due to this mechanism, the mountainous area gradually becomes lighter and the oceanic floor becomes heavier, and thus the state of balance or isostasy between these two areas gets disturbed but the balance has to be maintained.

In brief, when an uplifted land mass or mountain is worn down by the agents of erosion, the load on the underlying column of the crust is reduced by the weight of the material that has been eroded away. At the same time, a neighbouring column underlying a region of delta and sea floor where the denuded material is being deposited, receives a corresponding increase of load. At the base of the crust the pressure exerted by the unloaded column is increased. In response to this pressure difference in the mantle there is a subcrustal slow flow of sima from under the loaded column towards the base of the unloaded column. The loaded column, therefore, sinks, while the unloaded column rises. This process, whereby isostatic equilibrium is resorted, is called *isostatic adjustment.* This is made possible by a compensating transfer of material in the sub-crust. For example, extensive parts of North America and Eurasia were subsided under the enormous weight of accumulation of thick ice sheets during Pleistocene glaciation, but the land masses began to rise suddenly because of release of pressure of superincumbent thick load of ice sheets due to deglaciation and consequent melting of ice sheets about 25,000 years ago, and thus the isostatic balance was disturbed.

The raised beaches of Finland and Scandinavia show that an uplift of about 250 metres has already taken place during the last

8,000 years and the region is still out of isostatic balance and is rising.

The theories discussed above and the given examples illustrate several important facts about gravity and isostatic adjustments:

1. Gravity is the driving force for all isostatic adjustments. All types of loading and unloading, therefore, cause vertical movements. Isostasy is involved in all of the processes that shift material on earth's surface. Some of the more obvious isostatic adjustments to be expected are as follows:
   a. In mountains and highlands, as erosion removes material, the crust should rebound.
   b. In deltaic areas, where sediment is deposited, the added weight should cause the crust tosubside.
   c. In areas of volcanic activity, the added weight of excursions should cause the crust tosubside.
   d. In regions of continental glaciation, the thick ice sheet should cause the crust to subside. As the ice is removed, the crust should rebound.
2. Very small loads, such as water a few hundred metres deep, are also sufficient to cause isostatic adjustments.
3. Isostatic adjustments can occur very rapidly in a geological time frame (60 metres in fewer than 20,000 years).

Oceanic Basin

The ocean basins are composed of—the abyssal floor, technically inactive rises and seamounts. Broad, relatively smooth surfaces are known as abyssal floors. The abyssal floor extends from the continental margins to the oceanic ridges at a depth of about 3000 m. The abyssal floor consists of two sections—the abyssal plains and the abyssal hills.

Abyssal Plains: Abyssal plains are tectonically inactive areas of very less slope with gradient less than 1:1000. They have almost no relief as they are 1000 km wide and are situated at depths of 3-6 km. The flatness is due to the accumulation of fine sediments up to thickness of 1 km burying most relief features. The plains are covered at their centres by some 300 m of sediment which thickens towards the continental rise, slope and shelf. Generally, abyssal plains are located near the margins of continents where sediments from continental mass are transported by turbidity currents and spread over adjacent ocean floor.

Abyssal Hills: The abyssal hills are relatively small hills rising from 50 to 1000 m. They are up to 100 km wide and are circular or elliptical in outline. Abyssal hills appear to be associated with ocean ridges which are, in some places, parallel particularly in the Atlantic and occur in profusion in parts of the ocean floor separated from land by trenches. In the Pacific, abyssal hills cover between 80 and 85 per cent of the ocean floor.

Seamounts and Guyots: Seamounts and Guyots are isolated submarine volcanic features 2-100 km wide and more than 1000 m high whose surfaces lie less than 2000 in below sea level. Many seamounts and guyots, when they rise above sea level, form volcanic islands and among these, Hawaii and Azores are most notable. Many seamounts are sharply pointed but others are flat-topped hills whose summits lie more than 200 m below sea level. These flat topped seamounts are called guyots and there are about 1400 guyots in the Pacific alone.

Guyots have steep sides (12°-35°) and are considered to be levelled platforms, submerged largely by sea floor subsidence and also by the post-glacial rise in sea level. There is one strong evidence supporting this view. Guyots are commonly capped by coral growth in the tropics, which could not have been built at depths greater than 150 m.

Oceans' Ridge

The oceanic ridge is the most pronounced feature on the Earth covering nearly 23 per cent of the Earth's surface. This area is almost as large as the surface of the continents. An oceanic ridge is a broad transversely fractured linear swell lying mostly near the centre of oceans. The Mid Atlantic Ridge is the largest and the best known ridge whose highest part extends well above the sea level to form the Azores, Ascensian and other islands. Other well known ridges are the Carlsberg ridge in the Indian Ocean, the East Pacific ridge in the Pacific Ocean, the Lomonosov ridge in the Arctic Ocean and the Pacific Antarctic ridge between Antarctica, New Zealand and Australia. The broad characteristics of oceanic ridges are—

1. They are more than 1500 km wide and rise 1-3 km over the ocean floor.
2. They are continuous features around the entire globe extending from Arctic basin down through the centre of the Atlantic into the Indian Ocean and across the south Pacific before ending in the Gulf of California. The total length of the continental ridges is more than 84000 km.
3. A rift valley marks the crest of the ridges throughout their length.
4. The oceanic ridge, throughout most of its length, is cut by a series of transform faults.

The main difference between ridges and rises is that ridges are steep sided while rises, such as the East Pacific rise, are gentle sloped.

Ridges and rises can be both active and inactive. In active ridges and rises there is continued volcanic and seismic activity while inactive aseismic ridges and rises are devoid of any earthquake or volcanic activity. Ridges and rises are mostly arranged in the form of a linear chain of extinct volcanoes and are indicators of past plate movement. The most important aseismic systems are the Galpagos rise, the Rio Grande rise, the Hawaiian and the Austral Marshall Gilbert chain.

Oceanic Trench

Along some coasts, e.g., the western Pacific Ocean, eastern Indian Ocean and the Caribbean sea, the continental shelf drops steeply into trenches, forming the deepest part of the oceans. These trenches are 30-100 km wide and 300-5000 km long. Their sides slope first at angles of about 4-8° and then at 10°-16° to depths of more than 10,000 m. The walls are in the form of steps, which may have been formed by the slumping on young mountain belts on their seaward side. These are characteristically V-shaped with narrow flat floors up to several kilometres wide. They are covered by a very thin layer of sediments. Examples of trenches are—the Mariana trench of Philippines, the Java trench, the Aleutian trench, the Japan trench, the Puerto Rico trench, the Peru-Chile trench, etc.

Characteristics of Trenches

Trench (km)	*Depth (km)*	*Length Width (km)*	*Avg*
Pacific Ocean			
Kurile-Kamchatka Trench	10.5	2200	
Japan Trench	8.4	800	100
Bonin Trench	9.8	800	90
Mariana Trench	11.0	2550	70
Philippine Trench	10.5	1400	60
Tonga Trench	10.8	1400	55
Kermadoc Trench	10.0	1500	40
Aleutian Trench	7.7	3700	50
Middle America Trench	6.7	2800	40
Peru-Chile Trench	8.1	5900	100
Indian Ocean			
Java Trench	7.5	4500	80
Atlantic Ocean			
Puerto Rico Trench	8.6	1550	120
South Sandwich Trench	8.4	1450	90
Romanche Trench	7.9	300	60

Chain of Volcanic Islands

Island arcs are chains of volcanic islands. These islands are generally convex towards the open ocean (hence they are called arcs) and run parallel to the ocean trenches and mountain chains. Island arcs range in size from less than 1 km to as large as Guinea, Luzon or Hokkaido. These are seismically active and are topographically and structurally continuous with some continental belts of young folded mountains (e.g., Malaya, Kamchatka, Alaska), which they strongly resemble. Each island arc is generally made of the parallel arcs of island 50-150 km apart.

The outer arc is dominantly composed of folded and thrust sediments and lacks forms. It may be submerged, e.g., Ryukyu Islands between Japan and Taiwan or it may be topographically and structurally fused with the inner arc, e.g., Japan.

Troughs are simply depressions in the deep sea floor of which Barleltt trough is the most important example.

Marginal sea basins lie between the island arcs and the continents, e.g., the Philippine Sea. These basins are 500-1000 km wide and may include abyssal plains extending down to more than 5 km below sea level.

Description of Plateaus

Plateaus are submarine elevations of considerable extent with relatively flat tops. Examples of plateaus are the Albatros plateau in the Pacific Ocean off south and central America, the Seychelles plateau of the Indian Ocean and the Azores plateau in the north Atlantic Ocean.

Continental and Oceanic Crust

Thickness:—

1. Continental crust ranges in thickness from 30-50 km. Oceanic crust is 10-20 km thick.

Age

2. Continental crust contains the oldest rocks on the Earth, which ranges in age up to 3.8 b.y. Oceanic crust is less than 200 m.y. old.

Rock type

3. Continental crust is basically granitic in composition and also contains andesitic lavas, ashes, intrusive granites along with some basalt. Oceanic crust is composed largely of basalt and to some extent gabbro.
4. Continental crust contains a wide range of limestones, sandstones, shales and conglomerates. Oceanic crust contains calcareous and siliceous oozes, red clay and flash association.
5. Continental crust consists of a wide range of metamorphic rocks—gneiss, schist, slate, marble and metamorphic rocks compositionally similar to granite. Oceanic crust does not contain any metamorphic rock.

Deformation

6. Continental crust has been extremely deformed. The oceanic crust has not been deformed.

Processes

7. The internal energy sources on continents give rise to marginal volcanic activity, deep burial of rocks, metamorphism and granitisation. Internal energy sources result in the creation of ocean floor as well as subduction of the ocean floor.
8. External energy sources manifest themselves in subaerial weathering, erosion, transport, deposition by running water, ice, wind, and waves along the coasts of continents. External energy sources manifest themselves in waves, tides and currents in the oceans.

Relief

9. Continental positive relief features include fold mountains, uplifted plateau and volcanoes; plains include platform areas, low-lying shield, continental shelves and coastal plains. Oceanic positive relief features include ocean ridges, volcanoes and plains including abyssal plains.
10. Continental negative relief features include rift valleys, eroded valleys by streams and glaciers, and deflation hollows. Oceanic negative relief features include ocean trenches and submarine canyons.

Processes of Deformation

Rocks, whether they are igneous, sedimentary or metamorphic, can be subjected to powerful stress by tectonic forces, gravity and weight of overlying rocks. There are three types of stress:

1. Compression (shortening).
2. Tension (stretching).
3. Shearing (stress when two pieces slide past each other).

The strain that results from these stresses (how the rocks respond) is expressed as *folding* (bending) or *faulting* (breaking).

Rock Bending

A flexure or bending of the rocks of the earth's crust owing to compressional forces is known as *folding*. If the flexure takes the form of an arch, the fold is termed an *anticline* (more generally an *antiform),* while a flexure in the form of a trough is a *syncline* (more generally *synform).*

The line along which a change in the amount and/or direction of dip takes place is known as the *hinge line*, and on many folds this coincides with the position of maximum curvature. The area adjacent to the hinge line is known as the *hinge area* or nose of the fold. The limb of a fold is the part which lies between one hinge and the next, and the angle between the limbs is called the *inter-limb angle.*

A fold is said to close in the direction in which the limbs converge. As seen on a map, one says that a fold "closes to the northwest, etc." As seen in a vertical section, a fold may be said to close upwards (antiform) or downwards (synform). For an antiformal fold the direction of closure is the direction of plunge. A plane which joins the hinge lines of the successive beds in a fold is the axial plane. Fold axis is a term which has been used by different authors with different meanings, but from a practical point of view the best definition appears to be that it is a line parallel to the hinge. This may be expressed alternatively as the trace of the intersection of the axial plane on any bed constituting the fold.

It should be noted that the trace of the axial plane on the ground surface coincides with the axis only for folds of certain specific attitude. A *crest line* marks the highest points on the same bed in an antiformal fold, whereas a *trough line* marks the lowest points on the same bed in a formal fold. The crestal plane and trough plane are the planes connecting the crest lines and trough lines respectively of the successive beds of a fold. Axial planes, crestal planes and trough planes are not always flat, and may be curved in various ways, and also the axial plane of a fold is not necessarily coincident with the crestal plane.

Different Types

The type, shape and size of the folds depend upon the intensity and direction of the compressive forces and upon the nature of the rocks of the different strata of the sedimentary rocks. On the basis of shape and size, folds have been classified under the following categories: (i) symmetrical, (ii) asymmetrical, (iii) one limb-vertical, (iv) isoclinal, (v) recumbent, and (vi) anticlinal.

Symmetrical Fold (Simple, Upright or Open Fold): If the axial plane bisects the fold, the fold is said to be symmetrical or simple or open fold. In other words, where the compression is relatively mild or more or less equal from both sides, the folds are simple, open, upright or symmetrical. Simple or symmetrical folds are, however, rare. The Jura mountains of France and Switzerland are a typical example of symmetrical or simple folds. The Jura folds lie just to the north of the main collision orogen of the Alps and are therefore called as *foreland folds.* The Zagros mountains of southwest Iran are another example of foreland folds.

Asymmetrical Fold: If the axial plane has a dip, the fold is described as inclined or asymmetrical fold. The asymmetrical folds are characterised by unequal and irregular limbs. In asymmetrical folds, both the limbs are inclined at different angles. One limb is relatively larger and the inclination is moderate, while the other limb is relatively shorter with steep inclination and length.

One Limb-Vertical Fold: When the axial plane is inclined in such a way that one limb is vertical and the other steadily inclined, such a fold is called as one limb-vertical fold.

Isoclinal Fold: If the two limbs of a fold are parallel, the fold is said to be isoclinal. Isoclinal folds are formed when the compressive force is so strong that both the limbs of the fold become parallel but not horizontal.

Recumbent Fold: This is an overturned fold in which the axial plane is virtually horizontal.

Anticlinal Fold: An arched fold or upfold in the strata of the earth's crust. The two sides or limbs of the fold dip in opposite directions away from a crest-line or central axis. If the dip of each limb is equal, it is a symmetric anticline; if the angle of dip of one limb is greater than the other the anticline will be asymmetric (asymmetric fold) or overfolded. Because denudation generally attacks the arches of folded structures faster than it does the downfolds (syncline), it is common for these to become the location of valleys which run along the anticlinal axis, thereby exposing older rocks in the core of the anticline. In some areas, a complex anticline of great lateral extent is found in which many minor folds occur. Such a feature is known as *anticlinorium.*

Syncline: A syncline is a downwarped fold—just the reverse of an anticline. The sides of a syncline dip towards each other, meeting at the bottom of the fold. If synclines are closed due to the downward plunge on all sides of the rocks that make up the structure, the resulting form is basin or canoe-shaped. The sizes vary as greatly as in case of anticlines (synclines occur very commonly besides anticlines).

A complex syncline of great lateral extent in which many minor folds occur is known as *synclinorium.* Contrary to this, a series of smaller anticlines and synclines which form part of a larger arched structure in the earth's crust is known as anticlinorium.

Nappe: Nappe is a fold in which the axial plane is horizontal or sub-horizontal. It is a large scale tectonic over fold in the earth's crustal rocks, which has moved forward as a recumbent fold sometimes for tens of kilometres along a thrust plane (thrust-fault), thereby detaching itself from the so-called zone of roots. Nappes were first described in the Alps. A few examples of nappes have also been traced in the Himalayas. These nappes are in the Kashmir Himalaya (Wadia), Shimla Himalaya (Pilgrim), Garhwal Himalaya (Auden), and Kumaun Himalaya (Heim and Gansser).

The Failures

Faults are fractures in earth's crust along which slippage or displacement has occurred. A fracture in rock along which there has been an observable amount of displacement. In other words, fault is a rupture or fracture of rock strata due to strain, in which displacement is observable. Most faults occur in groupings, termed as *fault zone* or *fracture zone.* Thus fault zones are areas of crustal movement. At the moment of fracture the fault line shifts and a sharp release of energy occurs, called an earthquake or quake.

Faulting is accompanied by a displacement—a slipping motion—along the plane of breakage, or *fault plane.* Faults are often of great horizontal extent, so that the surface trace, or *fault line,* can sometimes be followed along the ground for many kilometres. Most major faults extend down into the crust for at least several kilometres.

Faulting occurs in sudden slippage movements that generate earthquakes. A single fault movement may result in slippage of as little as a centimetre or as much as 15 metres (about 50 feet). Successive movements may occur many years or decades apart, even several centuries apart. Over long time span, the accumulated displacement can amount to tens or hundreds of kilometres.

The following are the main nomenclatures of a fault:

Fault Plane: The plane along which the rocks are displaced by tensional and compressional forces acting vertically and horizontally is known as a *fault plane.* A fault plane may be vertical, horizontal, inclined, curved or any other form.

Fault Dip: The angle which the fault plane makes with the vertical plane is known as *fault dip.* In other words, it is the angle between the fault plane and horizontal plane.

Fault Strike: The term 'strike' is applied to a fault plane in precisely the same way as to bedding plane.

Upthrow Side: The uppermost block of a fault is known as upthrow side.

Downthrow Side: The lowermost block of a fault is known as *downthrow side.*

Hanging Wall: The upper wall of a fault is known as *hanging wall.*

Foot Wall: The lower wall of a fault is known as a *foot wall.*

Fault Scarp: A cliff formed directly by the displacement of a recent fault, but usually on a small scale (10 m in height). It is a relatively transient land form since denudation will soon modify the scarp, turning it into a fault line scarp or obliterating it altogether.

Several types of faults have been recognised by the geologists and geomorphologists. Some of the important faults are described as under:

Normal Fault (Tension Fault): The faults having primarily vertical movement are called *normal faults.* A normal fault occurs when rocks are pulled apart. The plane of slippage or the fault plane is steep (usually between 65 and 90 degrees), or nearly vertical. A normal fault results in a steep, straight, cliff-like feature called *fault scarp.* Fault scarp range in height from a few metres to a few hundred metres. Their length is usually measurable in kilometres. In some cases they attain lengths as great as over 300 km (about 200 miles). A narrow block dropped down between two normal faults is a *graben.* A narrow block elevated between two normal faults is a *horst.* Grabens make conspicuous topographic trenches, with straight, parallel walls, while horsts make block-like plateaus or mountains, often with a flat top but steep, straight sides.

Reverse Fault (Thrust Fault): *A reverse fault* is one in which the hanging wall is on the upthrow side. This is also called a *thrust fault.* On account of extreme compression rocks snap, and one stratum is pushed over the underlying stratum, i.e., the upper side is displaced above the fault plane relative to the side below. There is thus shortening of the crust in reverse or thrust fault as a result of compressional forces.

Lateral or Strike-Slip Fault (Tear Fault): A *strike-slip fault* is formed when the rock blocks are displaced horizontally along the fault plane due to horizontal movement. The best known of the strike-slip faults is the Great San Andreas fault of California, nearly 1,200 km long. Strike-slip faults result from shear stresses in the crust. They are commonly produced where one tectonic plate slides past another at a transform fault boundary. They also form as 'accommodation' faults, when two segments of the crust are stretched or shortened at different rates.

Step Fault: A normal fault which, when repeated by a series of parallel faults each with an increased throw in the same direction, will produce a stepped slope. Step faults are common in a fault-block topography and may be present on the flanks of a rift valley.

Hinge Fault: A hinge fault forms where the displacement increases from zero to a maximum along the strike.

Rift Valley: A linear depression or trough created by the sinking of the intermediate crustal rocks between two or more parallel faults is known as a 'graben' and the accompanying morphological feature as a *rift valley.* The East African Rift Valley, and the Rhine-Graben are some of the typical examples of these features. The Rift Valley of East Africa Rifting of continental lithosphere is the very first stage in the splitting apart of continent to form a new ocean basin. The process is beautifully illustrated by the East African Rift Valley System. This region has attracted the attention of geologists since the 1900s. They gave the name *rift valley* to what is basically a *graben,* but with a more complex history that includes the building of volcanoes on the graben floor.

The East African Rift Valley System extends from the Red Sea southward about 3,000 km (about 1,800 miles). Along this axis, the earth's crust is being lifted and spread apart in a long, ridgelike swell. The rift valley system consists of a number of graben-like troughs, each in a separate rift valley ranging in width from about 30 to 60 km (20 to 40 miles). Major rivers and several long, deep lakes, for example, Lake Nyasa and Lake Rudolph, occupy some of the valley floors. Two great volcanoes have been built close to the rift valley east of Lake Victoria. One is the Mount Kilimanjaro, whose summit rises to over 5,895 m (about 19,160 ft.). The other, Mount Kenya, is only a little lower and lies on the equator.

Importance of Failures

Faults have great geomorphological, ecological and socio-economic significance. They are the producers of earthquakes, and pulverisers of the rocks. From the economic point of view, petroleum deposits are found in porous sedimentary rocks that have been faulted against impervious shale beds. Consequently, areas of faulted sedimentary strata are favourite areas of oil exploration. Faults also give rise to the underground water table along fault planes which result into springs (hot and cold). In fact, most of the hot and cold springs are situated along fault lines. Moreover, fault scarps can form topographic barriers across which it is difficult to build roads and railways. They may also give rise to waterfalls.

Faults also have tremendous importance in the weathering of rocks. The courses of small streams often follow joint systems. While stream erosion, particularly on the outer bends of meandering rivers, is made more rapid and effective by joints. Glaciers and waves would be much less effective erosive agents, if they worked on perfectly solid rather than jointed rocks.

Isostasy Law

The state of balance which the earth's crust tends to maintain is known as *isostasy*. The word 'isostasy' has been derived from the Greek language. The gravitational adjustment of earth's crust is isostasy (Greek *isos*, 'equal'; *stasis*, 'standing'), a state of balance, i.e., if anything occurs to modify the existing state, a compensating change will occur to maintain a balance. The concept of isostasy is based on the principle of buoyancy, first outlined by Archimedes. Buoynacy is the ability of an object to float in a fluid by displacing a volume of that fluid equal in mass to the floating object's own mass. A steel ship floats because its shape displaces a volume of water equal in weight to its own weight plus the weight of its cargo. Thus, an empty containership displaces a smaller volume of water than the same ship when fully loaded. The water supporting the ship is not 'strong' in the mechanical sense, water does not support a ship in the same way a strong steel bridge supports the weight of your car. Buoyancy rather than mechanical strength, supports the ship and her cargo.

Any region of a continent that projects above sea level is supported in the same way. As an extreme example, consider the continent containing Mount Everest—the highest of the earth's mountains at 8.84 km (29,008 feet) above sea level. Mount Everest and its neighbouring peaks are not supported by the mechanical strength of materials within the earth, nothing on (or in) our world is that strong. The mountainous upper surface of the continent floats high above sea level because the lithosphere of which it is part sinks into the plastic asthenosphere until it has displaced a volume of asthenosphere equal in mass to its own mass. The continent's mountains rest at great height, in balance with their subterranean underpinnings but susceptible to rising or falling as erosion or crustal stresses dictate. Lower regions are supported by shallower roots. Like a ship floating in water, the entire continent stands in isostatic equilibrium.

It is best illustrated by a high mountain chain which rises above the surface of the earth but has to be compensated by deep 'roots' of sialic *(sial)* material which penetrates deeply into the underlying *sima*. Isostasy occurs because the crust is buoyed by the more dense mantle beneath it, and each portion of the crust displaces the mantle according to its thickness and density. Denser crustal material sinks deeper into the mantle than less dense crustal material. Alternatively, thicker crustal material will extend to greater depth. Isostatic adjustment in earth's crust can be compared to adjustments in a sheet of ice floating on a lake as you skate on it. The layer of ice bends down beneath you, displacing a volume of water with a weight equal to your weight. As you move ahead, the ice rebounds behind you, and the displaced water flows back.

As a result of isostatic adjustments, high mountains and plateaus having a great vertical thickness sink deeper into the mantle than do areas of low elevation. Any thickness change in an area of the crust, such as the removal of material by erosion or the addition of material by sedimentation, volcanic extrusion, or accumulation of large continental glaciers, causes an isostatic adjustment.

Base and Evolution of Surface

In the beginning, the surface of the Earth was very much like that of Mercury, Venus and other planetary bodies. The Earth's entire surface was littered with craters and the temperature was also very high. There were no continents or ocean basins, no land or sea, and no life. It was only at the end of the Precambrian that the planet actually had continents, ocean basins and life. The precise way in which this evolution took place is not known and several questions have to be answered in the context of the origin and evolution of the Earth's crust. These questions are—

1. What was the extent of the first crust, was it local or of worldwide scale?
2. What were the processes responsible for the growth of the crust and at what rate did the early crust grow?
3. What was the composition of the early crust?
4. When and how did the crust got separated into continental and oceanic crust?

The oldest rocks of the Earth's crust collected from Greenland, Australia and South Africa are continental rocks roughly 3.8 b.y. old. These are not a part of the primitive crust but possess an advanced continental composition. Most of these rocks are metamorphosed sedimentary rock. This means that:

1. The oceans had already formed by this time and
2. Sediments demonstrate that some high standing continental platforms evolved very early in the Earth's history.

The mineralogy and chemistry of some of these Archean metamorphic rocks show that they crystallised at about 550° to 8000° C at 5 to 8 kilobars at depths of 2.7 km. These rocks were also underlain by an additional 35 km of crust. Thus, some type of continental crust 50-70 km thick existed on the Earth by 3.8 b.y. ago. It is presumed that the earliest terrestrial crust may have formed between 4.2 to 4.5 b.y. and since then it has been partially recycled and the first stable crust did not form until about 4 b.y. ago. The extent of the first crust may be determined by comparison with the lunar crust and crust of other terrestrial planets. For example, the lunar highlands appear to represent the remnants of the early lunar crust (4.5 to 4.3 b.y. old) which covered almost the entire lunar surface. On Mercury and Mars, the widespread primitive crust has been preserved. If the history of the Earth is similar to these planets, which of course it is, then it is very likely that the Earth also had an early crust that covered its surface.

The origin of the crust can be explained in terms of three groups of theories—

1. Inhomogenous Earth accretion model.
2. Catastrophic model.
3. Non-catastrophic model.

Inhomogenous Model of Earth's Accretion: This model assumes that there was a hot solar nebula. The temperature in the nebula was constantly falling. With the fall in temperature, the various compounds of the solar nebula condensed and gathered, i.e., solid particles collided to form planetary bodies in the disc around the Sun. This progressive condensation and accretion resulted in planetary growth. The zoned structure of the Earth with an iron core surrounded by a silicate mantle, may have been produced by sequential condensation and accretion of the Earth from the cooling solar nebula.

The last compounds to condense produced a thin layer of planetary surface, rich in alkaline and other volatile elements, which formed the first crust.

However, if the crust evolved in this manner, i.e., by inhomogenous accretion, the non-volatile elements like uranium, thorium and Rare Earth Elements (REE) should have been concentrated in the core and lower mantle. Instead, they are concentrated in the crust. This concentration may have been the result of magmatic transfer from within the Earth, producing a crust of magmatic origin and concentrated REE in the crust.

Influence of Surface

The Earth has been subjected to extensive severe bombardment. The tremendous energy given off during large impacts may have produced a localised melting within the Earth and the magmas derived from these meltings may have solidified to form the first crust. When large bodies hit the Earth's surface, craters were formed and energy was transferred to the Earth. These craters were approximately 10 km deep and there was a sudden drop in pressure. This sudden release of pressure together with the production of a series of radiating fractures resulted in partial melting of the mantle beneath the crater. Erosion of the crater rim, and intrusion and extrusion of dominantly mafic magma from underneath filled the crater. Fractional crystallisation of mafic magmas produced granitic magma, which were intruded at shallow depths. The crater, which was filled by sediments and igneous rocks, rose isostatically to form a protocontinent. The impact of collision may also have initiated a convection cell beneath the protocontinent, thus, thickening the crust and causing it to grow by peripheral magmatic accretion. Such a collision may also have initiated the formation of a mantle plume, which rose and provided a possibility for the future growth of continental nuclei. Perhaps the Archean continental nuclei were produced over rising mantle plumes that were triggered by the impact on the Earth's surface. Following this analogy that impact necessarily leads to generation of magma, significant amount of magma should have been produced on the Moon. This is, however, not the case.

Process of Deformation

The crust of the earth is only about 60 km (36 miles) in thickness. It is, however, continuously affected by the crustal formation and deformation processes. The earth's continental crust is formed and deformed by the tectonic activity, driven by our planet's internal energy.

Crust Formation Method

Tectonic activity produces continental crust that is quite varied. Nevertheless, continental crust generally can be thought of in three categories:

1. Residual mountains and continental cores ('shields') that are inactive remnants of ancient tectonic activity.
2. Tectonic mountains and land forms, produced by active folding and faulting movements that deform the crust.
3. Volcanic features, formed by the surface accumulation of molten rock from eruptions of subsurface materials.

Thus, several processes operate in concert to produce the continental crust.

Shields of Continent

Generally old, low elevation heartland regions of continental crust are known as *continental shields.* These are the nucleus of crystalline rock on which the continent 'grows' with additions of other crust and sediments. The nucleus is the creton, or heartland-region, of the continental crust. Cretons generally have been eroded to a low elevation and relief and are old (most exceed two billion years in age, but all are Preçambrian, or older than 570 million years). Portions of creton are covered with layers of sedimentary rocks that are quite stable over time. An example of such a platform is the region that stretches from the Rockies to the Appalachians and northward into central Canada. The Baltic Shield, Angara Shield, Australian Shield, African Shield, Brazilian Shield and Guianan Shield are the other such examples. A region where a creton is exposed at the surface is called a *continental shield.*

The Making

Continental crust results from a complex process that involves seafloor spreading and formation of oceanic crust, its subduction, remelting, and subsequent rise of magma. To understand this process, begin with the magma that originates in the asthenosphere and wells up along the mid-ocean ridges. It is less than 50 per cent silica and is rich in iron and magnesium. This material rises at the spreading centres, cools to form ocean floor, spreads outward, and collides with continental crust. Being denser the oceanic crust plunges back into the mantle, and remelts. This magma then rises and cools, forming more continent.

Bodies of silica-rich magma may reach the surface in explosive volcanic eruptions, or they may stop short and become subsurface intrusive bodies in the crust, cooling slowly to form crystalline plutons such as batholiths.

Terranes: The migrating crustal pieces, dragged about by processes of mantle convection and plate tectonics are known as *terranes*. Displaced terranes are distinct in their history, composition, and structure from the continents that accept them. 'Terranes' should not be confused with *'terrain'* which refers to the topography of a tract of land.

The Applachian mountains, extending from Alabama to the Maritime provinces of Canada, possess bits of land once attached to ancient portions of Europe, Africa, South America, Antarctica and various oceanic islands. These discoveries, barely a decade old, demonstrate how continents are assembled.

The Surface

The Earth is composed of three layers – the crust, mantle and the core. The crust is the outermost layer, mantle the intermediate layer and core is the innermost layer. The crust of the Earth is seismically defined as the solid Earth above the Mohorovicic discontinuity or in short, simply Moho. The crust is thin, rocky veneer that constitutes the continents and the floors of the oceans, and is not a homogenous shell. There are two distinct kinds of crust, which, because of their distinctive compositions and physical properties, determine the very existence of separate continents and ocean basins.

The Character

The crust is very thin, averaging only about 33 km in thickness. Compared to its lateral extent, the thickness is only of a very minuscule proportion. When examined by geological and chemical techniques, the rocks exposed at the land surface show great heterogeneity and regional variation. For example, the younger margins of continents consist largely of sediments derived from the continued erosion of the continental surface and transported to the coast, where most of it is deposited in shallow water on the continental shelf. Such sediments may form accumulations many kilometres thick. On the other hand, the oldest continental regions, the Precambrian Shields are often dominated by igneous rocks such as granite or by highly metamorphosed rocks such as gneiss.

Even within a restricted region of a few square kilometres, great changes are likely to occur in chemical composition, e.g., from granite to basalt or gabbro or even to ultramafic rocks, such as peridotite. Among sedimentary rocks, the chemical differences may be even greater. Sedimentary processes are even more effective than igneous ones in causing the sorting out or separating from one another of chemical components.

The more recent calculations of the composition of the Earth's crust are based on estimates of the relative volumes of the different types of sedimentary, igneous and metamorphic rocks and the composition of each. It can be seen that the composition of the whole continental crust is intermediate between granite and basalt. Estimates of the composition of the granitic layer alone indicate that in fact this is not granitic, but is also intermediate in character.

Every rock can be assumed to contain each of the naturally occurring elements, though perhaps only at extremely low concentrations. Some of these abundant 'trace' elements are relatively uncommon in everyday use, while many of the metals of importance to society are extremely rare in nature.

The composition of some types of sedimentary and igneous rocks and of the continental crust might suggest at first sight that silicon is the dominant element in the Earth's crust. This is not true. If the analyses are recalculated to show the concentration of each element instead of its oxide, it is seen that oxygen is the major constituent (46.5%) of the Earth's crust, with silicon at second place. This superiority of oxygen is even more enhanced if the relative numbers of atoms of each element are calculated. Nearly two-thirds of the atoms of the Earth's crust are of oxygen.

On a volume basis, oxygen is 94 per cent of the Earth's crust. The relatively tiny silicon ions, each of which is only about a fortieth of the volume of an oxygen ion, make up less than 1 per cent of the volume of the crust. In effect, silicate minerals can be thought of as a stack of framework of oxygen ions, with the other much smaller ions fitting into the interstices and spaces between.

The densities of the crustal material are generally in the range of 2.5 to 3.3 g/cu cm. This makes them denser than water but less dense than materials beneath them. Thus, they would sink in water but float on the materials beneath it. The thickness of the crust varies and it can be known by the depth of the Moho, which is the deepest beneath regions of higher surface elevations. For example, the Moho reaches depths of about 70 km beneath large mountain ranges, 40 km beneath more common continental regions and only 6 km beneath the ocean bottom. Thus, the Earth's crust is composed of relatively light rigid materials than the denser, warmer, less rigid mantle materials beneath. In addition to the crust, the portion of the mantle just beneath the Moho is also rather cool and rigid as shown by seismic waves travelling through it. The upper cooler portion of the mantle, which encompasses the upper 80 to 100 km, sticks to the bottom of the crust and moves along with it. This combined rigid outer region is called the lithosphere (Greek: 'lithos -rock). Beneath the lithosphere, from 100 to 200 km depth, is the Low Velocity Zone (LVZ), which is much softer than the lithosphere. The prevalence of high temperature here means that some minerals are in a molten phase. This more plastic region of the mantle, which is largely composed of peridotite and on which lithosphere floats, is called the asthenosphere (Greek: *'asthenos'-weak)*.

The lithosphere is floating on the underlying asthenosphere. The massive portions of the lithosphere reach deeper to obtain the needed buoyancy. This floating lithosphere can be compared to a massive block of ice which extends deeper into the water while its top side extends proportionately higher into the air. Likewise, the lithosphere begins to sink in the asthenosphere when a weight is placed over it. Such an increase in the load may happen when a volcano grows or when the land is covered under a thick ice sheet. The adjustments which are required to obtain buoyant equilibrium is called isostatic adjustments and the condition of buoyant equilibrium is called isostasy. The condition of isostasy is best exhibited where massive portions of the lithosphere reach deeper into the asthenosphere to obtain the required buoyancy. Were it not for isostasy, mountain ranges would have gradually subsided because there are no rocks strong enough to bear the heavy load of mountain ranges. The mountains are not supported by the strength of the crust but rather are in a state of floatational equilibrium with the denser underlying rock.

Composition and Structure of Continental Crust: The Earth is the only terrestrial planet, which has a continental crust. Continents cover roughly about one-third of the planet. They are mostly concentrated in the Northern Hemisphere and are roughly triangular in shape. Continental crust range in thickness from 30 to 50 km. The depth of the Moho beneath the continents averages about 35 km, although the crust may be considerably thicker or thinner in particular regions. The thickest portion of the continents is found beneath mountain ranges. Continental crust has a density ranging from 2.5 to 2.7g/cu cm and this density conforms to the nature of rocks found on the continents. Although, the continental crust is referred to as being 'granitic', it is actually an assortment of various rocks which, if all melted together, could be converted to granites or more specifically granodiorite. Basically, the structure of continental crust comprises three layers—

1. *Upper Sedimentary Layer:* The continental sedimentary sequence rarely exceeds a few kilometres in thickness except in narrow, deeply subsiding tracts where the sediments accumulate to thicknesses in excess of 15 km. These sediments may be absent, particularly where the ancient magmatic and metamorphic rocks come out on to the surface, such as in the Baltic shield, the Aldan shield, the Anabar massif, etc.
2. *Granitic Layer:* Much of the continental crust is composed of metamorphic rocks of the same composition as granite. Continents mostly consist of a highly complex sequence of metamorphosed sediments and volcanic rocks, together with large volumes of granitic intrusions and granitic gneiss.
3. *Basaltic Layer:* Below the granitic layer is the basaltic layer. At some places the transition from granitic to basaltic layer is gradual but at some places, it is fairly clear (known as

Conrad discontinuity). This layer has a higher density compared to the upper layer.

One of the most important characteristics of igneous continental rocks is that they are relatively richer in silicon and potassium and poorer in iron, magnesium and calcium. These continental rocks are also the oldest rocks on the Earth and are as old as 3.8 billion years.

Geological Properties

In terms of basic geological character the continental crust can be divided into two types of regions—cratons and orogenic belts. Geological differences among continents are mostly in the size, shape and proportions of the two components.

The Cratons: Cratons (Greek: 'Aratos'-power, strength) are extensive flat stable regions of the continents in which complex crystalline rocks are exposed or buried beneath a relatively thin sedimentary cover. These regions have not been affected by any Earth movements for over a half billion years, except for broad gentle warping. These cratons include the shields, where large areas of highly deformed igneous and metamorphic rocks, the basement complexes are exposed. These stable platforms are stable regions of the crust covered by essentially horizontal sedimentary strata.

Shields: Shields are a regional surface of low relief having an elevation within a few hundred metres above sea level. Shields are broadly convex and relatively immobile regions, usually constructed of Precambrian metamorphic and igneous rocks.

The gentle, low relief is broken by resistant rock formations that rise a few tens of metres above the surrounding less resistant rocks. On a regional basis, shields are flat, slightly convex and almost featureless.

Shields are composed of a highly deformed sequence of Precambrian metamorphic rocks and granitic intrusions known as basement complexes. Most of the rocks are of the Precambrian age and are formed under conditions of high temperature and pressure, many kilometres below the surface. These rocks are traversed by faults and joints expressed at the surface by linear depressions.

The sedimentary volcanic rocks of the shield are highly deformed and as such has been converted to complex metamorphic rocks. These rocks were at a later stage intruded by granitic magmas. The upper cover of sedimentary and metamorphic terrain has been removed by erosion, exposing what we now see at the surface. These complex igneous and metamorphic rocks form the nucleus of every continent.

Shields of the World: Although, Precambrian exposures are common in cores of mountain ranges and canyons of plateaus, the most obvious and largest areas of Precambrian rocks are the erosionally stripped, regionally upwarped, geologically stable regions of the continents. These regions are called Precambrian shields because of their broadly convex shape (after Graeco-Roman shields). Although, shield rocks are exposed over only about 20 per cent of the Earth's land surface, they represent about 75 per cent of the observable time span accounted for by crustal rocks. Every continent has one or more Precambrian shields rimmed by Phanerozoic mobile belts.

The Indian Shield: The Indian shield is the most important geological region of India, apart from the Himalayan region and the Gangetic plain. There are a number of recognisable separate provinces in the Indian shield. The oldest is the Dharwar Province (2.4 b.y.), which acted as a nucleus around which other provinces became attached. It is predominantly granitic in character with intrusions of gneises and is exposed along the eastern side of the Indian Peninsula. Towards the northeast is the Eastern Ghats province, which is (1.6 b.y.) composed of similar rocks. The Satpura province, which lies at its northeastern edge stabilised about 1000 million years ago. Northeast of Satpura is the Aravalli province, again composed of crystalline rocks while the Vindhyan province contains stratified shales, sandstones and limestones resting unconformably over the Precambrian basement. Over 500,000 sq.km of the Indian Peninsula is covered by early Cretaceous-Cenozoic basalt.

The African Shield: Africa is basically a vast platform of abutting shield segments. Precambrian rocks are exposed over half the surface of Africa and elsewhere lie beneath the layer of Palaeozoic and Mesozoic rocks. It is a stable continent with only its eastern edge having been affected by epeirogenic movements producing faulting and fracturing, manifested in the African rift valleys.

The Early Precambrian events (before 2600 m.y) are found in the Berberton Mountain Land of South Africa, Transvaal and Rhodesian volcanic and granitic intrusion (2.6 and 3.1 b.y) that constitute the crystalline basement of South Africa. The Middle Precambrian, which began about 2.6 b.y ago and ended 1.1 to 1.8 b.y. ago, led to the evolution of four stable crustal segments. These are the West African bulge, two of which are in Central Africa and the remaining two are in South Africa. Quartzites, shales, conglomerates and lavas characterise the Middle Precambrian rocks. The Late Precambrian phase of Africa is characterised by the deposition of sedimentary rocks and widespread orogenies in the regions that bordered the old crustal segments. Strips of metamorphosed granitic crust formed between the segments welded them into a unified shield.

The South American Shields: There are three shields in South America—the Guianan, the Brazilian, and the Patagonian Shield. Gneises and schists form the oldest sequence, i.e., Early Precambrian. The Middle Precambrian consists of metamorphosed sediments, volcanic intrusions, and although metamorphic effects are less profound in the Late Precambrian, slates and phyllites are still evidenced along with quartzites, conglomerates, volcanic flows and ash beds. Because of the presence of a covering of younger sediments over parts of the shield as well as dense forests that cover many regions, the Precambrian rocks of South America are still little known in comparison to the relatively barren shields of Canada and Scandinavia.

The Angaran Shield: The Angaran shield is not exposed over a vast region, instead it is revealed as small exposed patches that elsewhere are covered by younger sedimentary rocks of the Siberian Platform. This shield forms the nucleus of Asia as it is around this shield that Asia has grown. The Himalayas and the Urals have actually got welded on to this shield. Like the Precambrian or other shields, the Angaran Shield too has an older complex of gneises, schists and granites and also a younger sequence of sedimentary rocks, volcanic and granitic intrusives that date back to about 1600 m.y. ago. These have been called Riphaean.

The most extensive exposure of Precambrian rocks, forms the surface of the Baltic shield. There have been four major episodes of mountain building in the Precambrian history of the shield. In each of these episodes, there were accumulations of great thickness of sediments, which were later subjected to severe compression, metamorphism and massive granitic intrusion. In the first two episodes, schists and gneises were formed followed by greenstones, slates, ripple marked sandstone and dolostones. In the third phase, the youngest consists of lavas and sandstones that were later intruded by granite.

The Canadian Shield: The Canadian Shield extends from the Arctic islands southward to the Great Lakes area and westward to the plains of western Canada. Most of the Canadian Shield is less than 300 m above sea level. The Canadian Shield represents the first stable and resistant granitic crust in North America. This shield was 20 to 60 per cent larger in the past than at present because the trends of the metamorphic structures were terminated abruptly by younger provinces. The rocks in the younger province were also metamorphosed but the original sediment was quite different from the sediment that formed in the earlier one. The concentric pattern of the provinces is a strong evidence that the continents grew by accretion of material around its margins.

The Australian Shield: Structurally Australia is composed of a vast Precambrian shield occupying most of the central and western parts of the continent. Mobile belts occur along the eastern part of the continent. As in other shields, the Precambrian surface is partially covered by younger sedimentary rocks. The most remarkable feature of the Australian Precambrian is the presence of thick sequences of relatively unaltered sedimentary rocks spanning over 1.5 b.y. The Australian shield comprises two regions— Archean group comprises igneous and metamorphic rocks over 2.5 b.y., represented by two episodes of mountain building. The Proterozoic rocks rests over worn roots of Archean Mountains. It consists of three great systems of quartzitic sandstones, basaltic lava flows and siliceous rocks.

The Antarctican Shield: Most of the eastern half of Antarctica consists of Precambrian shield. Archean rocks occupy the most extreme eastern edge while Proterozoics occupy remainder of the shield. Most of the western part of Antarctica consists of fold belts of Late Cretaceous and younger age.

The belts of metamorphic rocks together with igneous intrusions shows that shields are composed of a series of zones that were once mobile and tectonically active. Great mountain ranges once traversed parts of the shield but the mountains have since then got eroded and only the roots of these old ranges remain.

The Base

Platforms or stable platforms are those regions of the craton that are stable and consist of an underlying basement of ancient crystalline rocks covered by an essentially horizontal sedimentary strata. These areas have been relatively stable throughout the last 600-700 m.y., i.e., they have not been uplifted or submerged below sea level. The stable platforms form much of the broad flat lowlands of the world.

The rocks, although lying perfectly horizontal, locally seem to be warped into broad shallow domes and basins on a regional basis. The rocks, which cover the basement, are dominantly sandstone, shale and limestone, which were deposited in the ancient shallow seas and later re-emerged. This may have been by marine transgression and regression associated with plate movement.

Belts, which are Orogenic

Orogenic (Greek: *'oros* '-mountain, *'gen'*-production of) belts are the mobile belts arranged in linear or arcuate tracts that have been subjected to severe deformation and mountain building. These belts are typically found near the edges of the continents. Some mountain belts are still in the process of formation. Examples of such mountain belts are the Rockies, Andes and the Alpine-Himalayan chain (extending into Asia and northern Africa). In other mountain belts, deformation ceased long ago but there still exists considerable topographic relief. These include the Appalachian Mountains of the eastern United States of America, the Great Dividing Range of eastern Australia and the Ural Mountains of Europe.

There are two significant aspects related to the location of orogenic belts—

1. The young fold mountain belts are found along the continental margins. This implies that their formation is the result of the concentration of forces along the margins of continents and not from a uniform worldwide force evenly distributed over the Earth.
2. Many older mountain belts extend to the ocean and abruptly terminate at the continental margins. The northern Appalachian of the eastern United States, the Scottish Mountains and Atlas Mountains terminate at the continental margins. This suggests that these mountains were continuous long before and since then have got separated by continental rifting.

Description of Plains

Plains are flat areas with hardly any relief. They may be extensive, such as the Ganga, Brahmaputra, Amazon, Mississippi and Nile plains; linear or arcuate such as the coastal tracts; swampy, such as the deltaic plains of Sunderban, Mississippi, Orinocco, etc.; and, such as the desert plains of Mojave, Simpson, Sahara, Thar, Indus, etc. In fact, plains can be of an amazing variety and they can be classified according to shape, geographical location or origin. Genetically, plains can be either of volcanic origin, degradational or aggradational, fluvial, glacial, arid, marine or karst.

Structure and Composition of Oceanic Crust: The oceanic crust is approximately 5 to 12 km thick and has an average density of 3.0g/cu cm. The most important characteristic of the structure of the oceanic crust is its remarkable uniformity. There is a surface layer of marine sediments overlying igneous rocks, which form three distinct layers.

The upper sedimentary layer (Layer 1) varies considerably in thickness and composition. This layer consists of calcareous and siliceous shells of microscopic marine organisms together with red clay. The three igneous layers are each of uniform composition and thickness. The upper layer (Layer 2) consists of basaltic flows, which are 1 to 2.5 km thick. These basaltic flows are pillow shaped and were originally fed by numerous fissures. Later these basaltic flows solidified into dykes in the vents and fissures. The pillow lava represents the eruption of volcanoes on the sea floor. Layer 3 almost entirely consists of dykes.

Layer 4 consists of coarse-grained gabbro. Below the gabbro are peridotites composed almost entirely of olivines and pyroxenes. This material is considered to be a part of the mantle. The boundary between the crust and the mantle is called the Moho. Oceanic crust has a concentration of relatively heavier common elements such as iron magnesium and calcium.

Ocean Bottom Relief

The average depth of the seas is almost 3790 m compared with the average elevation of the land, that is, just 875 m. The highest peak on the continents, Mount Everest has an elevation of 8848 m, compared with a depth of almost 12,000 m for the greatest oceanic deep, the Mariana trench of the Pacific Ocean. In all, the depths of the seas are so much greater than the elevation of the continents that the average of the entire Earth's surface is almost 2430 m below sea level. The entire vertical distribution sequence can be better understood by means of a hypsographic or hypsometric curve.

The major elements of the ocean floor are—

Continental margins

- Continental shelf
- Continental slope
- Continental rise

Ocean Basin Floors

- Abyssal floor
- Seamounts and Guyots
- Inactive rise

Ocean ridges and rises

Ocean Trenches

Island arcs

Marginal ocean basins

Plateaus

Ocean-ridge flank	500-1500 km	<1 km	>3km
Ocean-ridge crest	500-1000 km	<2 km	2-4 km

Continental Margins: Continental margin comprises of—Continental Shelf, Slope and Rise.

Shelf of Continent

Geologically speaking, the continental shelf is not a part of the oceanic crust. This is because the continental shelf is composed of continental crust and sediment derived from the erosion of relief features on the land. The continental shelves are covered by the oceans and therefore, form a part of the ocean bottom relief. The continental shelf can be called a submerged part of the shield or the stable platform. The continental shelf has an average width of 78 km and an average maximum depth of 133 m. The depth ranges from 20-550 m at the edge of the continents and the width ranges from zero to almost 1500 km. The continental shelf is smoother than the flattest surfaces exposed on land since the average gradient is only 0.1 degree.

The continental shelf is smooth and flat but a detailed examination of the shelf does reveal some relief. On about 60 per cent of the surface, there are low hills with a relief of 18 m or more, 35 per cent of the shelf surface is marked by shallow basins or valleys having a relief of more than 18 m.

Only 30 per cent of the present shelf area is covered with recent marine beds. Two per cent of the area of the continental shelf is covered by terrestrial deposits laid down about 25,000 to 10,000 years ago on the river and coastal plains. This was because the greater part of the continental shelves was dry land during the glacio-eustatic marine regression in the last glaciation. At that time, broad expanses of lowland favoured the development of flood plains, delta plains and marshlands.

Slope of Continent

Like the continental shelf, the continental slope is also not a part of the oceanic crust. Continental slopes are basically continental margins covered by the ocean. The continental slope descends from the outer edge of the continental shelf as a long continuous slope to the ocean basins. The continental slope marks the edge of the continental granitic rock mass and thus, defines the boundary between continental and oceanic crust. The continental slope is basically the topographic expression of the geologic difference between the continental crust and the oceanic crust reflecting a fundamental difference in the structure and rock type.

Continental slopes are by far, the longest and the highest slopes on the Earth. The slope itself may be straight or smoothly curved or regionally interrupted by one or more platforms. Generalised slopes near the continental break range from 1 to 10°, averaging about 4.3°. The general gradient decreases as the slope merges with the deep sea floor at depths of about 3000-4500 m. The continental slope is steepest where mountain ranges border the coast (4-6°) and where faults demarcate the shoreline (5-6°). Continental slopes are very gentle opposite major deltas (1.3°) and stable coasts (2-3.5°).

The surface of the continental slope is not smooth. Numerous submarine canyons and furrows result in a large relief, and in many places the slope is marked by hills and ridges.

Submarine Canyons of the World

Pacific Ocean	*Atlantic Ocean*	*Indian Ocean*
Tokyo Canyon	Oceanographer Canyon	Indus Canyon
Bering Canyon	Hudson Canyon	Ganges Canyon
Columbia (or Astoria) Canyon	Wilmington Canyon	
Juan de Fucca	Norfolk Canyon	
Monterey Canyon	Congo Canyon	
Arguello Canyon	Sao Francisco Canyon	
Scripps Canyon	Mississippi Canyon	

In places especially opposite large rivers, such as the Congo and the Hudson, the continental shelf and slope are cut by submarine canyons-exhibiting seaward slopes, bifurcation and sinuosities analogous to subaerial rivers. The steep sided canyons are of many sizes (width 1-15 km, lengths 20-2000 km, gradients 1:40) and are partly drowned features resulting from erosion by density or turbidity currents. Turbidity currents lead to the formation of deep sea fans. These fans are located at the base of continental slopes with their apex at the mouth of a submarine canyon cut into the edge of the shelf. The chief source of sediment for the formation of the fan is mud, which is brought by rivers. These sediments are intermittently flushed through the canyon leading to the formation of fans where the currents reach the lower gradients of the ocean floor.

Many fans have deep channels, which are an extension of the submarine canyons cut in the continental slope. The channels develop natural levees analogous to stream natural levees on land. The greatest fans are found beyond the Ganges and the Indus delta. Other large fans are the Amazon and the Congo fans in the south Atlantic, Mississippi fan in the Gulf of Mexico and the Laurentian fan in the north Atlantic. Large fans are absent in the Pacific because most of the important rivers drain into the Atlantic and Indian oceans, and the deep marine trenches trap most of the sediment that flows into the Pacific.

Continental Rise: At the base of continental slopes, the slope decreases to 1° or less continuing into the abyssal hill or plains. From this decrease in slope onwards begins the continental rise. This is also a part of the continental crust.

Setup of Mechanism

Coverings of Earth

The Earth can be classified on the basis of either its chemical composition or mechanical properties.

Layers of Differing Chemical Composition: Three compositional layers are recognised on the Earth—core, the innermost layer, mantle, the intermediate layer and crust, the outermost layer.

At the centre of the Earth is the solid inner core. It is about the size of the Moon but has approximately the density of pure iron or an iron-nickel alloy. It is generally thought to have frozen out of the liquid outer core. The outer core is molten but is probably close to the freezing point. The inner core is, therefore, suspended in a relatively low-viscosity fluid and is fairly isolated from the rest of the Earth.

The outer core extends from the boundary of the inner core (5149 km depth) to the core-mantle boundary which is at depth of 2891 km, about half of the radius of the Earth. The outer core does not transmit shear waves and is, therefore, a fluid. The density of the core is about twice the density of the mantle and this also contributes to the velocity drop. Seismic waves entering the core from the mantle therefore refract downwards.

Internal zones of the Earth.

Zones		Depth (and radius) of boundaries in km		Velocity of P waves km/s	Density gm/cm^3	Pressure in units of 10^6 bar*
		Sea-level	(6371)		2.7	
CRUST (continental)†	A				2.8	
					2.9	
Mohorovicić Discontinuity		33	(6338)			9000
				7.9–8.1	3.32	
		50	Low velo-	7.8		
		250	city zone	8.1		
	B					
UPPER MANTLE		400	(5971)	8.90 9.13	3.54 3.72	140 000 270 000
	C	(720: deepest earthquakes)				
		670	(5701)	10.27 10.75	3.99 4.38	382 000
LOWER MANTLE	D					
Core/Mantle Boundary		2891	(3480)	13.71 8.06	5.57 9.90	1 368 000
OUTER CORE	E/F					
		5149	(1222)	10.36 11.03	12.2 12.8	c.3 300 000
INNER CORE	G					
		6371	(Centre)		13.1	c.3 600 000

* Standard pressure of 1 atmosphere = 1 031 250 dyne/cm^2
1 bar = 1 000 000 dyne/cm^2

† In oceanic areas Zone A extends from the ocean floor at a mean depth of 4.8 km to the Moho at a depth of about 10–11 km below

It has been concluded that the outer core is not much more viscous than water, and may even be more than twenty orders of magnitude less viscous than the mantle. It therefore converts readily and cannot support large internal temperature or chemical gradients. The outer core is probably the most homogenous part of the Earth but this does not rule out light layers at the top and dense layers at the bottom where material may have settled out. In fact, the inner core may have formed from iron particles settling through the outer core. The core seems to have cooled over time and as it cools, crystals of pure iron may form from the iron-oxygen (and/or sulphur) mixture, which has a lower freezing temperature. Production of these crystals will induce a form of chemical convection and this may help stir the core and provide some of the energy to drive the terrestrial 'dynamo' and generate the Earth's magnetic field.

Seismic waves reflect well off the core-mantle boundaries (CMB), good evidence that the boundary is sharp and exhibits a large change of velocity and density over a fraction of a seismic wavelength. There is an irregular solid layer just above the CMB, which has a thickness of 200 to 300 km. This is called the "D" layer. This layer is variable in both velocity and thickness and in some places is separated from the overlying mantle by a sharp but small discontinuity.

In some places the shear velocity gradient in this region is negative, i.e., the velocity *decreases* with depth. These features are best explained by the existence of a chemically distinct layer at the base of the mantle and this layer is also a thermal boundary layer. The core is convecting and is losing heat to the overlying cooler mantle.

The CMB represents the largest density contrast in the Earth. It is therefore a natural collection point for any light material leaving the core of dense material settling out of the mantle.

The lower mantle is the largest subdivision of the Earth. It extends from the top of "D" (about 2740 km depth) to the major seismic discontinuity at a depth near 650 km. The lower mantle is 49.2 per cent by mass, of the Earth and 72.9 per cent of the mantle plus crust. The predominant mineral in the lower mantle is having the composition (MgFe) SiO_3.

The second most abundant mineral is probably (Mg, FeO) in the NaCl or rock-salt structure.

The next important feature of the mantle is the 650 km discontinuity (sometimes called the 670 km discontinuity, but it may vary in depth from place to place by 50 to 100 km).

The region of the mantle between the 650 km discontinuity and another mantle discontinuity at 400 km is known as the transition region, being the transition between the upper and lower mantle.

In the upper mantle, down to about 400 km depth, the seismic velocities are consistent with a mixture of olivine and orthopyroxene (peridotite), or garnet and clinnopyroxene (eclogite), or a mixture of peridotite and eclogite (so-called fertile peridotite). The deepest samples, from kimberlite pipes, come from as deep as 200 km and these are primarily peridotite although eclogites are not uncommon.

Basalt is the predominant material that reaches the surface from the interior of the Earth. A variety of evidence suggests that basalts separate from their immediate parent at depths shallower than about 100 km, but the ultimate source region may be much deeper, perhaps in the transition region. Buoyant upwelling in the mantle is probably responsible for much of the lateral heterogeneity of the upper mantle, which is being mapped by seismologists using the methods of seismic topography.

Mechanical Partition

Lithosphere: Because they are so close to the cool outer surface of the Earth, the upper 80 to 100 kilometres of the upper mantle also are rather cool and rigid like the crust and move along with the crust as a unit. This combined cool rigid outer region, which includes both the crust and top layer of the upper mantle, is called the lithosphere. Although we use considerable vertical exaggeration in our drawings, the lithosphere is still extremely thin in comparison to its width. Its relative dimensions are thinner than those of an egg shell, for example. The lithosphere is divided into several large fragments called plates, and the motion of these surface plates over the years is referred to as plate tectonics. The lithosphere appears to be quite complex, especially beneath the continents, where it sometimes includes fragments of former crustal plates that have been thrust downward during collisions.

Asthenosphere: Immediately beneath the lithosphere, at a depth extending from about 100 to 200 kilometres, is the low velocity zone. The name is derived from the relatively slow speed of seismic waves travelling through this region. Compared to the lithosphere above, this region is more plastic or less viscous, that is, it softer, more pliable, and capable of bending or deforming without breaking. The higher temperatures at these depths cause some partial melting of the materials, which means that some materials melt or soften and lubricate the motion of others that are more rigid or resilient. Beneath the low velocity zone is no sudden or decisive change in physical properties like occurs at the top where it abuts the lithosphere. The pressure increases with depth, which causes the speed of seismic waves to increase as well.

The entire relatively plastic region of the upper mantle, including the low velocity zone and some of the material beneath it is called the asthenosphere. Scientists do not completely agree on the lower boundary of this region, but many now view it as extending down to the 670-kilometre-deep base of the upper mantle. From this viewpoint, we could divide the upper mantle into two regions the upper 80 to 100 kilometres of cool rigid material than together with the crust make the lithosphere, and the lower 570 to 590 kilometres of warmer and more plastic material that make up the asthenosphere.

It is important to remember that descriptive words like "soft", "plastic", or "movement" are written from a geological perspective, not a human one. Although the asthenosphere is soft compared to the lithosphere, it is still much more rigid than steel, for example. Furthermore, the movement we refer to is on a time scale of millions of years, not human lifetimes. Things that bend or flow over millions of years can be quite resilient to the blow of a hammer.

Mesosphere: Mesosphere includes the earth's zone that lies below the asthenosphere and includes the whole of mantle.

Barysphere: The term now less used includes the whole of core.

Pressure of Earth

The pressure deep in the earth is reasonably well known, but the temperature is less well known. The pressures are due to the weight of the overlying material, and, with our knowledge of specific gravity gained from seismic studies, can be calculated from the law of gravity. These pressures are hydrostatic; that is, they are equal in all directions because the rocks in the mantle and crust are not, as we will see, completely rigid, even though they are solid. A simple analogy will illustrate hydrostatic pressure. Bricks piled one on the other exert a pressure only on their bases, because they are rigid. A pile of sand, however, unless confined by a wall, will assume a cone shape because the sand is not rigid; hence, the pressure due to its weight is exerted in all directions. Data from drill holes and deep mines reveal that the temperature rises with depth in the earth at the approximate rate of 1°C per 27 metres (1°F per 50 feet). If this rise continues to the centre of the earth, the temperature there would be 222,000°C (400,000°F), an unreasonably high temperature—hotter than the sun. The temperature in the earth is estimated from knowledge of the melting points of the materials assumed to form each layer, taking into account the rise in melting point at high pressures.

Important Terms

- Anisotropic medium
- Asthenosphere
- Barysphere
- Conrad discontinuity
- Core
- Crust
- Free earth oscillations
- Gutenberg—Weichert discontinuity
- Isotropic medium
- Lithosphere
- Love wave
- Mantle
- Mesosphere
- Meteorites
- Mohorovicic discontinuity
- Primary wave
- Rayleigh wave
- Reflection
- Refraction
- Rigidity modulus
- Secondary wave
- Sesmic waves
- Shadow zone
- Snell's Law
- Thermal Boundary Layer
- Wave channel

Earth's Interior

The interior of the earth and its composition has always remained a matter of great controversy among the geologists and geophysicists. At present, a lot of reliable knowledge is available about the composition and structure of the earth. In fast the surface of the earth has been adequately sampled and analysed. However, the deepest hole in the earth (a drill hole) is only about 8 km deep, and there is no method of sampling from greater depth. Since there is no way of finding the chemical composition of the unseen or inaccessible material, ideas on the composition of deeper regions must be based on indirect evidence, and therefore are somewhat speculative.

In the absence of direct evidence, the physical properties of the inaccessible regions of the earth can be measured or estimated more readily. The change in temperature, pressure and density with depth can be estimated from the velocity of earthquake waves through the earth. The thermal and physical state of the interior of the earth also helps in ascertaining the structure and composition of the earth.

Earth's Temperature

It is an established fact that as we go down from the surface, inside the earth, the temperature goes on increasing at the rate of 1°C for every 32 metres of descent. At this rate the temperature at the depth of 48 km (asthenosphere) would be between 1,200°C and 2,000°C. The rocks and minerals cannot remain in a solid state at such a high temperature. It is, therefore, inferred that the source of lava eruption of the volcanoes is at a depth of about 48 km (30 miles). If we assume that the temperature goes on increasing at this rate (1°C after every 32 metres), the temperature at the centre of the earth should be more than 4,000°C. It has also been concluded by the earth scientists that the rate of increase of temperature inside the earth undergoes a sharp decrease at greater depths from the surface. In the upper layers of the earth an important source of the heat is radioactive minerals like uranium and thorium. These minerals are more abundant up to the depth of about 100 km, below which they are found in much lesser quantities. It, thus, appears that the rate of increase in temperature decreases downwards with increasing depth. The thermal conditions of the interior of the earth may be summarised as under:

1. At a depth of about 48 km the temperature is about 1,100°C.
2. The temperature at the depths of 400 km and 700 km is about 1,500°C and 1,900°C respectively.
3. The temperature at the junction of the mantle and the outer molten core (i.e., at a depth of 2,900 km) is about 3,700°C.
4. The temperature at the junction of the outer molten core and the inner solid core (i.e., at a depth of 5,100 km) is about 4,300°C.

The pressure exerted by the weight of the atmosphere inside the earth has also been used to determine the composition and structure of the interior of the earth. One atmospheric unit equals to a pressure of about 14.7 lb per sq inch. At this rate, at a depth of 2,500 km (1,500 miles), the pressure is about one million atmospheres and in the centre of the earth the pressure is estimated to be about 3.5 million atmospheres. Such a high pressure in the interior of the earth exercises a powerful influence on the temperature and physical state of the earth.

Normal Density

The average density of the earth is 5.5. The surface layer of the continents, composed of *sial* (silica+ aluminium) or granite rocks, however, has a density of about 2.7. From these facts it may be inferred that the density of materials in the core of the earth is much more than the density of rocks found in the uppermost layer. The material in the core of the earth is probably 10 to 12 times heavier than water. It is believed that the core of the earth consists of *nife* (iron + nickel).

The density of the intermediate layers is about 4.3. It consists of *sima* (silica+magnesium).

The geologists have consensus on the point that the earth is composed of several shells or layers, the density of which goes on increasing with depth.

Advanced Waves

The behaviour of seismic waves in the different layers of the earth provides the most authentic evidence about the composition and structure of the earth. The different types of waves, generated during the occurrence of an earthquake, are generally divided into three broad categories: (i) primary waves, (ii) secondary waves, and (iii) surface waves. Seismic waves—both P-waves and S-waves—travel faster through rigid material than through soft or plastic material. The velocities of these waves travelling through a specific part of earth thus give an indication of the type of rock there. Abrupt changes in the seismic wave velocities indicate significant changes in earth's interior.

Primary Waves: These are also called as longitudinal or compressional waves. In this type of wave motion, particles of the medium vibrate along the direction of propagation of the wave. These are the high frequency, short wavelength, longitudinal waves. These waves travel not only through the solid part of the earth but also through the liquid part of the core. A primary wave travels with fastest speed through solid materials, and under certain circumstances, it can change into a secondary wave on refraction, or vice versa. In the liquid materials, their speed is slowed down. These waves are analogous to sound waves wherein particles move both to and fro in the line of propagation of the ray. Much of our knowledge of the internal structure of the earth has been gained from the study of primary and secondary waves.

Secondary Waves: These are also called as transverse or distortional waves. Secondary waves are analogous to water ripples or light waves, wherein the particles move at right angles to the rays. A secondary wave cannot pass through liquid materials. These are the high frequency, short wavelength waves which are propagated in all directions from the focus and travel at varying velocities (proportional to density) through the solid part of the earth's crust, mantle and core.

Surface Waves: Surface waves are called as long-period waves. These waves generally affect the surface of the earth only and die out at smaller depth.

The surface waves are characterised with low frequency, long wavelength, and transverse vibrations which develop in the immediate neighbourhood of the *epicentre.* The surface waves are responsible for most of the destructive force of earthquakes. They are confined to the outer skin of the crust. They cover the longest distance of all the seismic waves.

When an earthquake occurs the seismic waves are recorded at the epicentre with the help of seismograph. In the beginning a few small and weak swings are recorded. Such tremors are called as 'the first preliminary tremors'. After a brief interval 'the second preliminary tremors' are recorded and finally 'the main tremors' of strong waves are recorded.

Seismic Waves as Probes of Earth's Interior: Seismic waves passing through the earth are refracted in ways that show distinct discontinuities within earth's interior and provide the basis for the belief that earth has: (1) a solid inner core, (2) a liquid outer core, (3) a soft asthenosphere, and (4) a rigid lithosphere.

As stated above, the seismic waves are similar in many respects to light waves, and their paths are governed by laws similar to those of optics. Both seismic vibrations and light vibrations move in a straight line through a homogeneous body. If they encounter a boundary between different substances, however, they are reflected or refracted. Familiar examples are light waves reflected from a mirror or refracted (bent) as they pass from air to water.

If the earth were a homogeneous solid, seismic waves would travel through it at a constant speed. A *seismic ray* (a line perpendicular to the wave front) would then be a straight line. Early investigations, however, found that seismic waves arrive progressively sooner than was expected at stations progressively farther from an earthquake's source. The rays arriving at a distant station travel deeper through the earth than those reaching stations closer to the epicentre. Obviously, then, if travel times of long-distance waves are progressively shortened as they go deeper into earth, they must travel more rapidly at depth than they do near the surface. The significant conclusion drawn from these facts is that the earth is not a homogeneous, uniform mass but has physical properties that change with depth. As a result, seismic rays are believed to follow curved paths through the earth.

In 1906, scientists recognised that whenever an earthquake occurs, there is large region on the opposite side of the earth where the seismic waves are not detectable. To better understand the nature and significance of this *shadow zone*.

For an earthquake at a particular spot (labelled '0°'), the shadow zone for P-waves invariably exists between 103 and 143 degrees from the earthquake focus. Evidently, something deflects the waves from a linear path. The best explanation for this shadow zone is that earth has a core through which P-waves travel relatively slowly. Seismic rays through the mantle follow a curved path from the earthquake's focus and emerge at the surface between zero and 103 degrees from the focus (slightly more than a quarter of the distance around the earth).

In 1914, Beno Gutenberg, a German seismologist, calculated that the depth to the surface of the core is 2,900 km. Later analysis of more numerous and more reliable seismic data showed that Gutenberg's original estimate was remarkably accurate, with a probable error of less than two-thirds of 1 per cent. This suggests the presence of a solid inner core, which deflects the deep, penetrating P-waves.

The core has even more pronounced effect on S-waves, but this effect cannot be explained by reflection or refraction. S-waves simply do not pass through the core. There is a huge shadow zone extending almost half way around the earth, opposite the earthquake's focus. One difference between P-waves and S-waves is particularly significant.

P-waves pass through any substance— solid, liquid, or gas. S-waves, however, are transmitted only through solids that have enough elastic strength to return to their former shapes after being distorted by the wave motion. They cannot be transmitted through a liquid. The fact that S-waves will not travel through the core, therefore, is generally taken to mean that the outer core is liquid.

The Gaps

Seismologists have located within the earth two major layers which separate S zones within the earth having markedly different properties. The outer one—the *Mohorovicic Discontinuity*—separates the crust from the mantle, its average depth being 35 km. Separating the mantle from the core is the *Gutenberg Discontinuity* at about 2,900 km.

There, S-waves stop, and the velocities of P-waves are drastically reduced. The outer core is thought to be made of molten iron mixed with some light elements like Si, S, O, C, or H. The extra component is needed to explain the density of the core, about 10 per cent less than pure iron.

Other discontinuities are apparent, but less striking. The first occurs between 5 and 70 km below the surface. This is known as the *Mohorovicic Discontinuity* or simply *Moho,* after Andrija Mohorovicic, the Yugoslavian seismologist who first recognised it. The discontinuity is considered to represent the base of the crust and heralds an important compositional change from the feldspar-rich *(sial)* crust to the olivine-rich *(sima)* mantle. Seismic studies show that the continental crust is much thicker (25 to 70 km) than oceanic crust (about 8 km).

Perhaps, the most significant discontinuity, however, is the low-velocity zone from 100 to 250 km below the surface. The normal trend is for seismic wave velocities to increase with depth in the mantle. In this low-velocity zone, however, the trends is reversed, and seismic waves travel about 6 per cent more slowly than they travel in adjacent regions. The generally accepted explanation for the low seismic-wave velocities in this zone is that the material is partly molten, perhaps no more than 1-10 per cent. The low-velocity zone is included within the globe-enriching *asthenosphere.* This weak layer exists because the mantle is near its melting point. The asthenosphere plays a key role in the motion of tectonic plates at the surface of the earth.

Table below delineates conditions immediately adjacent to the discontinuities, not average uniform conditions (for P-waves and S-waves).

Seismic waves (Velocities in km/sec)

Velocity of P-waves			6.5	16.6	
Velocity of S-waves			3.74	7.4	
Density		Crust	2.95	5.7	Mantle
	Above				
Discontinuity			*Mohorovicic*		*Gutenberg*
	Below				
Velocity of P-waves			3.76	8.1	
Velocity of S-waves			4.36	Not transmitted	
Density		Mantle	3.3-3.5	9.5	Core

Source: Whitten (1977:130).

Seismic Imaging

The seismic tomography, which is like its medical analogy CAT Scan (computer assisted tomography: tomograph is based on a Greek word, *tomos,* meaning section), also proves the given structure and composition of the earth.

In seismic tomography, seismic waves are used like x-rays. Seismologists analyse the velocity of hundreds of thousands of seismic waves as they pass through the earth in different directions. The results are images showing regions where the waves travel faster or slower than normal. Geophysicists and geomorphologists come across from laboratory studies as well as from observations near volcanoes that seismic waves are slowed by unusually hot rocks and are speeded up by cooler rocks. The tomography thus can be interpreted as 'temperature maps'. Hot parts of the mantle, being less dense than their surroundings, will rise, where as cool mantle rocks will sink. Thus, the tomograph can be used to outline the pattern of flow in the mantle.

The three-dimensional view of the mantle obtained from seismic tomography confirms what we might expect from the plate tectonics theory. At a depth of 150 km slow seismic zones occur under most of the volcanic regions including the mid-ocean ridges. In contrast, the shields of Canada, Brazil, Siberia, Africa and Australia are all fast. At a depth of 350 km the mid-oceanic ridge system is no longer continuous but is broken up into isolated segments. At a depth of 500 km there is even less of a relationship between mantle and surface features. This indicates that the mid-ocean ridge system is not simply the surface expression of vertical upwelling currents from the deepest mantle. Instead, it must be fed by the lateral transport of hot material in the upper mantle.

Seismic tomography is also providing a more sophisticated view of the core. Early maps of the core-mantle boundary show that the surface of the core is not smooth, but is marked by swells and depressions with a difference in height up to 20 km. The presence of any kind of topography on the core-mantle boundary bears on a number of fundamental geologic questions. A rough boundary would presumably disturb the flow of the liquid iron in the outer core, much as mountain influences the flow pattern of winds. The topography of the core also has important implications for the flow of energy in the earth. A rough core boundary would imply that a lot of energy escapes the core as heat which might be the source of heat for convection in the mantle and the resulting tectonic plates.

Thus, our newly acquired ability to construct three-dimensional tomographic images of the earth's deep interior opens up the breathtaking prospects of tracing tectonic plates as they plunge below the surface in a subduction zone and descend into the deep mantle.

The interchanges among new tools, new observations, and scientific theories are nicely exemplified by these advances in understanding the earth. Progress in our ability to use remote imaging (seismic tomography) for the distribution of temperature in the mantle has led to new hypotheses about mantle convection patterns. These observations have forced other scientists to re-examine old theories and make new computations to test the validity of suggestions.

The resulting computer models have reproduced the distribution and sizes of some observed cold blobs and hot plumes and they have led to predictions about mantle flow patterns. Now, new observations fit the theory.

Many other basic geologic questions must look to the earth's interior for answers. Projects planned for the future include increasing the number of seismic stations so that we might obtain sharper global image of the mantle and the hidden flow that shapes the surface of the earth.

On the basis of the study of the seismic waves and seismic tomography, it may be concluded that the earth has several density zones inside. The behaviour of seismic waves and the tomographic analysis proves beyond doubt that the earth consists of three layers: (i) the crust, (ii) the mantle, and (iii) the core.

Levels of Earth

The Crust: The outer layers of the earth's structure, varying between 6 km and 48 km in thickness, and comprising all the material above the Mohorovicic Discontinuity. It is divided into two shells, a lower, continuous, layer—the *sima* (acronym of silica and magnesia)—and an upper, discontinuous, layer—the *sial* (acronym of silica and aluminium). The sial is apparently confined to the continental masses.

Velocity of earthquake waves in sial and sima

	Sial	*Sima*
Thickness	10-12 km	15-20 km
Density	2.7	2.95
Velocity of earthquake waves	P 5.57 km/sec	6.5 km/sec
S 3.36 km/sec	3.74 km/sec	
Composition	Granite	Basalt

Source: Whitten (1977:109).

The crust, as a whole, is thickest beneath the mountains and thinnest under the oceans. It is composed of various kinds of rocks. In its uppermost part we find sedimentary rocks. This sedimentary layer is not continuous over the entire surface of the earth and is generally thin. The thickness of the sedimentary rocks is less than 3.2 km but in areas of folded mountains this may increase to 32 km or more. Below the sedimentary cover is a layer of crystalline rocks, consisting of granite and gneises in its upper section and basaltic rocks in the lower section. Sometimes these crystalline rocks cover wide areas on the surface of the earth, such as the Western Australia, Peninsular India, Middle Africa, Brazilian Plateau, Eastern Canada, Scandinavia and North-East Asia.

The contact zone of the crust and the mantle is called *Mohorovicic* or *Moho Discontinuity*. Here, the rocks are different in chemical composition from those below and above. The depth of the *Moho* varies from 5 to 7 km beneath the oceanic plains, and 45 to 70 km from the surface in the folded mountain areas.

The Mantle: The layer of the earth between the crust and the core, with its upper boundary marked by the *Mohorovicic Discontinuity* and its lower boundary by the *Gutenberg Discontinuity*. The depth of mantle varies between 35 km and 2,900 km. The density ranges from 3.3 at the Mohorovicic Discontinuity to 5.7 at the Gutenberg Discontinuity. The mantle is composed of dense and rigid rocks which have predominance of minerals like magnesium and iron *(olivine)*, hence the term *peridotite shell* which is sometimes used. It has been suggested that there are substantial quantities of sulphides in the upper mantle and some nickel-iron in the lower mantle. It has been claimed that minor discontinuities can be detected within the mantle.

The mantle may be divided into two parts: (i) the lower mantle or the *mesosphere;* and (ii) the upper mantle or the *asthenosphere.* The lower part of the asthenosphere like the mesosphere is solid, but the upper part of the asthenosphere is plastic and is in a partially molten condition. The velocity of the earthquake waves decreases in the asthenosphere, and it is, therefore, referred as the 'lower velocity zone'.

The Core: The central part of the earth is known as core. It probably consists of a dense nickel-iron alloy *(nife)* with a temperature estimated at about our 2,700°C. The outer perimeter of the core commences at the Gutenberg Discontinuity, 2,900 km from the surface, where the outer core may be liquid. Within this appears a much denser inner core with a radius of about 1,400 km which may be solid. The density of the core ranges from 9.5 to 14.5 and sometimes even higher.

The S-waves, which can pass through only solid objects, suddenly disappear at a depth of 2,900 km. Further, at this depth, the velocity of the P-waves, which can travel through solid, liquid and gases, abruptly decreases from 13.7 km to 8 km per second. This has been identified as the outer limit of the core.

To conclude, the velocities of P-waves and S-waves through the earth indicate that the earth has a solid inner core, a liquid outer core, a thick mantle, a soft asthenosphere, and a rigid lithosphere.

Earth—Interior Study

A study of the interior of the Earth is necessary to understand many of the fundamental mechanisms underlying the Earth's dynamic surface. The surface of the Earth is readily sampled and analysed and therefore, there is a good deal of information about it. However, the factual evidence concerning the interior of the Earth is not readily available. Geological process can expose rocks created at some 20-25 km depth and volcanoes erupt pieces of rock that were once part of the Earth's upper mantle. The deepest hole in the Earth—a drill hole is only about 8 km deep and there is no method of sampling at greater depth. Apart from these data there is no direct evidence concerning the interior of the Earth. For these reasons, ideas on the interior of the Earth are based on indirect evidences and are therefore somewhat speculative.

Exploring the Interior

The study of the Earth's interior can be approached in two ways:

1. By considering the Earth as part of the solar system and deducing indirectly such factors as its speed of rotation, density and mass (all of which fall within the domain of astronomy).
2. By analysing direct information, particularly that provided by seismic waves produced by natural or artificial earthquakes, and by the measurement of gravity.

Density Study

Density is defined as the mass per unit volume and is usually expressed in grams per cubic centimetre written as g/cu cm. If the earth is considered to be spherical in shape, then using the known value of the Earth's radius and mass, its density can be determined. The average density of the Earth is thus calculated to be 5.52 g/cu cm. This calculated density is considerably greater than the average density of the earth's crust (2.8 g/cu cm). In order that the average of the earth's density has such a high value, the material below the crust must have a density much greater than that of the crust. Since the Earth wobbles only very slightly as it rotates and since the acceleration due to gravity over the Earth's surface is quite uniform, the earth's mass must be distributed uniformly about the Earth's centre as in a series of concentric layers. A study of density, gravity and the Earth's dimension makes it possible to calculate the pressures within the Earth and also the temperatures that can be reached under this pressure.

The Meteorites: Meteorites are considered to be the remnants of unknown planets. Some meteorites are made of stony silicate or rocky lumps, some composed mainly of iron, nickel and other metals, a few are stony iron with metal inclusions. Meteorites allow us to directly analyse the density, chemistry and mineralogy of the nickel iron cores of bodies having a similar composition to that of the Earth.

The Seismic Studies: Seismic study involves the study of the seismic waves. Seismic waves are vibrations usually created by disturbances within the earth's crust. These disturbances may be due to slippages along an earthquake fault or an underground nuclear explosion. Seismic waves travel through the Earth and are detected at seismograph stations on the Earth's surface. A seismic activity generates two types of waves— (a) body waves and (b) surface waves.

The body waves are of two types:

1. ***P Waves** or **Primary Waves** or **Compressional Waves:*** In these waves the individual particles of the material, in which the waves are travelling, oscillate back and forth in the direction of propagation.
2. ***S Waves or Secondary Waves or Transverse or Shear Waves:*** In these waves the oscillations occur transverse to the direction of propagation.

Surface waves are confined to the surface of the earth. They are of two types Love waves and Rayleigh Waves.

A wave is the propagation of strain through a material. If any small volume within an elastic medium is distorted or strained in some way, a stress is developed, which tends to restore the material to its original undisturbed state. The ratio of stress to strain is the elastic modulus of the material. Because a material can be strained in different ways so it has more than one wave velocity. The general expression for the wave velocity of two seismic waves are:

$Vp = V(K+4u/3/d)$

$Vs = Vu7d$ where,

k = bulk modulus (a measure of the stress needed to compress the material to a smaller volume)

p = rigidity modulus (a measure of the stress needed to change the shape of the material)

sd = density

P waves both compress and change the shape of the material and hence depend on density and compressibility of the material. S waves depend on density and rigidity of the material and therefore cannot propagate through a liquid, which cannot resist change in shape and hence has no rigidity.

Seismic waves have two properties, which make them particularly useful in revealing the Earth's internal structure.

1. Earthquake waves are reflected when they strike the interface between two different materials having different elastic properties and density. This reflection is in accordance with Snell's law.
2. Seismic waves tend to bend or refract to when they enter a medium in which they travel at a slower speed. This is not a unique property of seismic waves rather, it is shared by all waves.

When a wave strikes an interface between two materials, the amount by which it is bent depends on the difference in wave velocities between the two materials. The greater the change in speed, the greater will be the refraction of the waves. When the wave speed changes gradually within a material the wave will also bend gradually. The more gradual the change in wave speed the more gradually the wave will bend. Consequently, when a seismic wave crosses an interface between two distinct regions the change will be abrupt and when it passes through regions where there are gradual changes in pressure, temperature or composition the waves will bend gradually.

These seismic waves can be detected at various other points on the surface. The time required for the wave to arrive at any detector depends on the path of the wave through the Earth and on the speed of the wave along that path and the two are related. Since the path followed by the wave depends on the distribution of wave speeds within the Earth, therefore from an analysis of arrival times for seismic waves at various points along the Earth's surface we can use the known relationship between wave speeds and paths to discover exactly how the speed of seismic waves varies with depth inside the Earth. If the wave speed is known at any given depth, this can provide physical parameters such as density, rigidity and compressibility for the various layers within the major structural units of the Earth. The paths taken by the seismic waves (body waves) as they pass through the Earth provide information about the dimensions, structure and physical properties of each of the internal layers. If the interior of the Earth had uniform properties (it was an isotrcpic medium) the waves would follow straight lines and their speed would not change but when they pass through an anisotropic medium they follow a curved path. The earthquake waves are reflected and refracted when they come to boundaries of different elastic properties or densities (discontinuity) and follow a curved path with change in velocity. These discontinuities are due to changes in composition, atomic structure or atomic state and occur at several places within the Earth.

As a single medium cannot transmit waves of the same type (P or S) at different speeds, when two compression waves are received separately at a station, the implication is that they must have been propagated in different media (wave channels). The discovery of these channels and their transitional zone is an important seismological result.

Seismographic Study

For any given earthquake, the seismographs obtained from different stations at increasing distances from the epicentre are examined and it is found that

1. From 0 to 200 km, there is a compression wave propagated at a more or less constant speed of 5.6 km/sec^{-1} and a shear wave at 3.4 km sec^{-1}. These are said to be individual waves and are written as Pg and Sg.
2. In 1909, A. Mohorovicic found two sets of P and S waves. There was a difference in the travel times of the first and second impulses of P and also in those of S. The Pg (5.6 km/sec^{-1}) attenuated progressively, disappearing completely at a distance of 200 km. from the epicentre. The normal wave P, however, could still be found at a distance of 11500 km from the focus.

Moreover further from the epicentre reflections were much smaller. Mohorovicic concluded that the additional impulses had travelled by some route other than the direct path from the epicentre to the recorder. This means that there was a discontinuity at a depth of some 50 km separating a superficial medium in which the waves moved at low speed from a deeper one in which wave speeds were much higher. The P and S waves travelling in a deeper layer, on reaching the upper boundary of this layer, are refracted along it and move through the lower layer with a higher velocity than those Pg and Sg following a direct path. This discontinuity otherwise known as Mohorovicic discontinuity or Moho is the boundary between the crust and the mantle. On a global scale the Moho is shallow under the oceans (some dozen kilometres), deep under mountain chains and intermediate under eroded land masses.

3. Conrad found two wave channels above the Moho. The upper channel had waves at 5.6 km sec^{-1} (Pg) and 3.36 km sec^{-1} (Sg). The wave channel in the lower channel however was P_b = 6.3 km sec^{-1} and S_b = 3.6 km sec^{-1}. The implication is that the crust itself is divided into two parts-the upper crust separated by a discontinuity (Conrad discontinuity at 20 km) from the lower crust.
4. From 800 km to 11500 km (=103°) the Pg and P_b waves are progressively dampened so that only the normal waves (P) reach the stations.
5. From 11500 km to 14000 km (=142°), no waves are received directly by the stations, this is the shadow zone.

Beyond 142° the P waves are found to appear again in a modified form and without the S vibrations. The slowing down of the P waves and the complete elimination of S waves can be interpreted thus.

The slowing down of the P waves was due to their having passed through matter of different composition from that near the surface. The material evidently lay at greater depth than that penetrated by the waves received at 103° and is due to the presence of core of different composition within the Earth. Further, the loss of S waves means that they must have passed through material without any rigidity and hence became extinct (since S waves cannot be transmitted through liquid, which has no rigidity). The increase in P wave velocity and its slowing down and refraction enables the depth of the material to be computed. By determining where S waves are and are not received on the far side from a disturbance, the dimension of the liquid material (the core) can be mapped. Thus, at 2900 km a discontinuity exists, known as the Gutenberg-Wiechert discontinuity, which defines the mantle-core boundary.

Thus, on the basis of seismic investigation, the Earth can be divided into three major zones.

1. The outer superficial layer called crust, extends down to 30 to 40 km beneath the continents and even further beneath some mountain regions and to about 10 km beneath sea level in oceanic regions. At the base of the crust, earthquake waves increase abruptly in velocity as they enter a denser layer, the mantle. This boundary is called the Moho or M discontinuity.
2. The mantle extends from the Moho to 2890 km. The mantle is a solid layer and is itself divided into three layers - an upper mantle from Moho to 400 km, an intermediate mantle from 400 to 1000 km and a lower mantle from 1000 km to 2890 km.
3. At the base of the mantle, there is an abrupt reduction in the velocity of P waves and the disappearance of S waves. This discontinuity known as the Gutenberg-Wiechert discontinuity defines the core mantle boundary. The core-is a liquid at least in its outer part. The inner core (5150 to 6370 km) is solid and is composed largely of nickel and iron.

Additional information on the physical properties of the Earth's mantle and core may be obtained from an analysis of the free earth oscillations of the Earth. In P and S and ordinary seismic wave transmission, the motion is looked upon as a series of travelling disturbances which affect only a relatively small part of the Earth at any given time, without reference to what is happening at that time to the whole Earth. In the Chilean earthquake of May 1960, in addition to the regular P, S and surface waves, free earth oscillations were recorded for the first time in history which set the whole Earth vibrating as a single unit like a bell for as long as several weeks. The tone of the Earth's vibration is pitched too low for the human ear to hear but modern seismographs are sensitive enough to detect these low frequency oscillations.

There are two classes of such oscillations, torsional (or toroidal) and spheroidal. In the torsional mode, there is no displacement at the equator and the two hemispheres oscillate in antiphase. The fundamental spheroidal oscillation is an alternating compression and rarefaction of the whole Earth. The next higher mode is the three zones. As the sphere oscillates it is distorted alternately into an oblate and prolate spheroid. Its period is 53 minutes.

There are higher modes with an increasing number of subdivided zonal distributions and each of these modes also has overtones with internal nodal surfaces corresponding to each fundamental. Each oscillation period observed, supplies a value which must be fitted by acceptable distributions of the Earth's density, compressibility and rigidity. Moreover, whereas the immediate data from P and S waves yield only the quotients of the elastic moduli and density, the free earth oscillations are not tied to these quotients.

The free oscillations excited by a major quake last for several days but their amplitude diminishes because the Earth is not a perfectly elastic body. The precise way in which these oscillations diminish can be determined for each mode of oscillation. From these observations information on the elastic properties of the Earth's interior can be obtained.

Formation of Earth

The origin of the earth is a fascinating subject in which the geophysicists, geologists and experts of earth sciences are increasingly interested. A number of theories have been put forward by them, but the subsequent discovery of facts about the heavenly bodies and other physical discoveries disclose the weakness of the various theories. It is said that out of all the subjects of physical geography the *'origin of earth'* is the most speculative, largely based on reasonable conjectures. Consequently, all the hypotheses and theories, developed about the origin of the earth, the solar system and the universe, have been subsequently rejected. It is most probable that we have much more to learn before a definite and lasting theory of the origin of the earth and the solar system is devised.

This subject, therefore, is one under exploration and investigation. As one rejects the past theories in the light of newly discovered physical facts, those theories which are deemed as modern may be ridiculed in future after the discovery of newer facts.

While attempting to understand the origin of the earth we deal with vast space, long time, high temperature and unknown gravitational fields. Looking at the complexities of the mathematical calculations involved, can we ever hope to get a satisfactory theory for the earth's origin in view of these difficulties? No one knows the answer to this question.

Modern Hypotheses

The German physicist Carl von Weizsacker showed in 1943, that all the old objections against the Kant and Laplace hypotheses can be easily removed.

This he could do on the basis of new knowledge about the chemical composition of the cosmic matter. Such knowledge made it clear that the chemical composition of the earth was different from that of the stellar bodies. The proportion of light gases like helium and hydrogen is very small on the earth. As against this, the 'elaborate spectroscopic analysis' showed that the sun was composed mostly of hydrogen and helium and the common constituents of the earth formed only about 1 per cent of the total mass of the sun.

It is now also known that space is not empty. It contains extremely diffuse and fine gas and dust, at the rate of about 1 milligram per 4 million cubic kilometre of space. Chemically, this matter is the same that constitutes the stars and the sun. The dust, however, forms about 1 per cent of the interstellar matter, the rest being mostly hydrogen and probably helium. This inference about the cosmic matter is based on the selective absorption of light from stars. Before reaching earth, the light of the stars takes thousands of light years.

In the words of Gamov, the new knowledge concerning the chemical constitution of matter in the universe plays directly into the hands of the Kant-Laplace hypotheses.

According to Weizsacker, when the sun was originally formed from the condensation of cosmic matter, an envelope of such matter, about hundred times the total mass of planets, remained outside the sun as a 'rotating envelope. The envelope had a diameter about as large as that of the solar system. The thickness of the disc was about 300 to 500 million kilometres. This envelope consisted of such gases as hydrogen and helium and such dust particles as iron oxides, compounds of silicon, water droplets and ice crystals.

The collision of dust particles and their accumulation into' larger aggregates resulted into planets.

Such collisions in the rotating envelope could be of the same force as that of meteorites. The aggregation of planetary matter is believed to have resulted not from the collision of equal particles which would have resulted in mutual pulverisation. Instead, collision occurring between larger particles and smaller ones would lead to the accretion of the latter to the former. Later, the gravitational attraction of the larger bodies on the smaller particles would cause further aggregation.

According to Weizsacker, the time of the formation of planets from the fine cosmic dust was about 100 million years. During the period of collision and constant bombardment caused by the accretion of cosmic matter, the planets remained hot but thereafter cooling caused the formation of solid crust.

The planets are so arranged that the distance of a planet from the sun is roughly twice the distance of the next inner planet. This is true of the eight planets, namely Mercury, Venus, Earth, Mars, Jupiter, Saturn, Uranus and Neptune, and also asteroids. Pluto is an exception. Its distance is only about one-third as large as that of Neptune from the sun.

A similar rule is found with regard to the distance of satellites from their planet, e.g., in the case of the satellites of Saturn.

Now, it needs to be explained why the envelope of rotating matter round the sun did not aggregate into one lump instead of several planets. It is believed that the size of particles forming interstellar space was about 0.0001 centimetre, and there were 'some 10^{45} particles moving along elliptical orbits of all sorts of sizes and elongations'. In such enormous traffic of particles collisions were certain. Such collisions either pulverised them or diverted them into less crowded zones. Some of the orbits were circular, others were elliptical.

Weizsacker showed that the particles were moving round the sun in a large number of bean-shaped 'necklaces' to ensure non-intersecting traffic pattern. The total picture was that of '8'. Collision occurred only in the boundary regions between such rings.

Such arrangement was conceived because it would avoid collision within each ring. As, however, the periods of rotation of the individual rings were different, collisions did take place in the marginal areas of the rings. From such collisions in the boundary regions the larger aggregations at specific distances from the sun occurred.

According to this picture, it could be shown that the radii of the boundary lines between the adjoining rings occur in geometrical progression. Each radius is about twice the preceding radius.

The same rule holds good for the satellites of the planets suggesting that the satellites originated more or less in the same way as the planets.

Of the envelope surrounding the sun, the gases formed about 99 per cent. The dust particles were aggregated into larger lumps to form the planets and their satellites, and the asteroids, but the gases, mostly hydrogen and helium, escaped into interstellar space in a period of some 100 million years.

Weizsacker's theory envisages that what happened to the sun must have occurred in the case of other stars also and numerous systems of planets must have formed.

Thus, there must be millions of planets in our galactic system alone with physical conditions resembling those of the earth. No wonder, therefore, that life "in its highest forms" existed in the planets of our galaxy.

The most inhabitable planets of the solar system are Mars and Venus. We may know in future about their habitability by means of spaceships.

The asteroids are apparently the source of meteorites. The former are only about 1 per cent of the mass of the earth.

Role of Meteorites

The meteorites are of four principal types:

1. The irons or iron meteorites composed of nickel-iron, nickel being about 7 per cent.
2. Stony irons. Here, in addition to nickel-iron, there are 'stones' of silicate minerals.
3. Chondrites which contain globular nodules composed of silicate minerals without or with glass. Some small quantity of nickel-iron is also present.

4. Achondrites are such meteorites which have no chondritic nodules nor nickel-iron. They approximate to basaltic composition and contain plagioclase felspar.

The meteorites contain some diamond, both in the iron and stony meteorites. The age of the meteorites, on the basis of "lead from stone meteorites", is believed to be about 4,550 million years.

The Criticisms

1. Weizsacker's theory has aroused much interest on the origin of the solar system but the details of his model have not been accepted. 'Turbulent eddies' to which the formation of planets is related, are not so regular in size and arrangement as was supposed by Weizsacker.
2. The passing of the sun into a cloud of dust and gas or 'diffuse' nebula could not be regarded as a rare event. The great nebula in the constellation—Orion—is a conspicuous nebula. There are several luminous as well as dark nebulae in the galaxy. The dark ones are those which have a blanketing effect on the distant stars. The nebulae are very extensive, so that if the sun happened to plunge itself into one of them, it would remain there for lakhs of years. The gravitational power of the sun would gather an envelope of interstellar matter. This, therefore, is in favour of Weizsacker's theory.

The theory "accounts satisfactorily for the present distances of the planets from the sun". We may consider one more theory of the nebular type.

Electromagnetic Law

In 1942 Hannes Alfven assumed that in the past the sun was rotating faster than now. It then happened to plunge itself into a nebula in which the atoms were originally electrically neutral. As the atoms were pulled by the gravitational attraction of the sun, they were ionised or electrically charged. Such unionisation extended in the sun's envelope of atoms upto planetary distances.

Then, Alfven applied the behaviour of electrically-charged particles in a magnetic field and showed that matter would gather largely in the equatorial plane of the sun at such distances which can now be compared with the distances of Jupiter and Saturn.

This matter would revolve round the sun and one of its effects would be to retard the rotation of the sun.

Later, the gaseous and other atoms condensed into large planets. It is believed that Jupiter having its own magnetic field behaved in the same way as the sun with regard to the surrounding matter, and Jupiter's satellites were formed.

The magnetic field of the sun is the basis of this theory. The electromagnetic force is far more powerful than its gravitational force. The ratio of former to the latter on a proton in the orbit of the earth is 60,000:1. It is 250:1 on a proton in Pluto's orbit.

But there are lacunae in the theory. One of these is that it does not explain the origin of the inner planets (Mars, Earth, Venus and Mercury).

Cataclysmic Law

Most of such theories depend on collision between two stars or their mutual approach. But the stars are mutually so immensely distant that the possibility of such a catastrophe is extremely remote.

As such the cataclysmic theories mostly have historical value and are indicative of the lines of thinking on the origin of the solar system in the light of the knowledge which was available at the time of propounding a particular theory and with a view to meeting objections against pre-existing theories.

Theory of Tides

Planetesimal Hypothesis of Chamberlain and Moulton: In 1904, Chamberlain and Moulton suggested the origin of the planets as due to severe tidal eruptions and disruption of some of the sun's mass. The former was caused by the gravitational power of an approaching star. The disrupted solar matter was dragged to immense distances from the sun. The larger parts of the tidal ejection gathered together. These relatively large bodies gathered smaller scattered bodies or planetesimals and eventually grew into the mature planets of the system. The main merits of the hypothesis when it was presented were as follows:

1. There was no need of assuming a molten earth.
2. The earth grew bit by bit by accretion of planetesimal matter.

3. When the mass of the earth became condensed and compacted, it developed internal heat. This created pockets of melting. Subsequently, there was differentiation of a metallic core and a stony outer crust.
4. The constituents of the atmosphere and oceans also originated from the planetesimals.
5. The basis of the hypothesis is the passing star close to the sun at a sufficiently high speed. In order that matter may be drawn out from the sun, the passing star must have come within a distance of the sun not more than the diameters of the two stars themselves. This seems to be an extremely unlikely event. Further, their closing velocity must have been at least 5,000 km (3,000 miles) per second. This figure is considered unreasonable because it exceeds the escape velocity of the galaxy.
6. The hypothesis has been built on a catastrophic and extraordinary event. It is unreasonable to base any concrete scientific hypothesis on such an event.
7. If the planets have been formed by planetesimal accretion, then they should possess much less angular momentum than they actually have.
8. According to Jeffreys, the nucleus of the planet was incapable of attracting the atmospheric elements. Even if there were some gases and water in the nucleus, it was not possible for them to come out once they had aggregated with the planetesimals. Thus, in his view, the hypothesis presents a wrong picture of the origin of the atmosphere and is unable to solve the other problems relating to the origin of the earth.

Theories of Jeans and Jeffreys: The British scientist Sir James Jeans advocated the tidal hypothesis about the origin of the solar system in 1919, and Sir Harold Jeffreys introduced some important modifications in it in 1926. The names of both Jeans and Jeffreys are, therefore, associated with the tidal hypothesis.

The tidal hypothesis about the origin of the earth and the solar system was developed as a reaction to the Laplacian hypothesis and the shortcomings of the planetesimal hypothesis of Chamberlain and Moulton.

According to the tidal hypothesis, the solar system has been formed as a result of the coming together of the sun and another passing star. A huge star, many times bigger than the sun, happened to pass close to the sun. The gravitational attraction of the passing star had a powerful tidal effect on the sun and through tidal action a part of the sun was raised in the form of a long cigar-like gaseous filament, which ultimately broke loose and separated from the sun on account of the high gravitational attraction of the star. The outer fragments of this filament were lost in space, while its inner parts fell back upon the surface of the sun, and the middle part condensed and compacted into sections like the beads of a garland to form the nine planets which started revolving round the sun.

The filament, according to Jeans, was unstable lengthwise and soon broke up into several sections, each forming a distinct aggregation of matter, later to develop by cooling and contraction into a planet. Jeffreys suggested, however, that the aggregations may come into being shortly after the ejection of the corresponding solar material. In other words, the filament may not be strictly regarded as a continuous but rather as a series of condensations whose masses and spatial distributions are shown by him to be roughly in accordance with the present position and characteristics of the planets in our solar system.

By a repetition of the same process, under the influence of the gravitational attraction of the sun and possibly also of the passing star, some materials further got separated from the planets and condensed to form the satellites. In general, the size of the planet determined the number of the satellites.

By the time of the origin of the satellites, the parent planets had been cooled to a liquid state. Thus, the satellites were born as liquid masses. It is believed that the original orbits of the planet were eccentric, not circular. But, the tidal effect of the passing star would have been to scatter solar matter round the sun in addition to the filament. The resistance offered by such scattered matter would reduce the eccentricity of the planetary orbits and make them circular.

Criticisms: The tidal hypothesis of Jeans and Jeffreys has been criticised on several counts as under:

1. Levin (1958) put forward the view that the stars lie at such long distances from each other in the space that the possibility of their coming close together or their mutual encounter can easily be ruled out.
2. Where did the big star that was responsible for tidal ejection of solar material go? No satisfactory answer is available to this question.
3. The gaseous matter which would be ejected from the sun would be so hot and there would be so much momentum in the gaseous particles that they would soon be lost in space.
4. If the planets have been formed from solar material, then the sun and the different planets should all be composed of similar materials. This is, however, not so. There is the predominance of lighter elements (hydrogen and helium) in the sun, while the planets are composed mostly of heavier elements like iron, silicon, aluminium, etc.
5. The hypothesis does not explain the distances between the sun and the different planets.

Binary Star Hypothesis of Russell: The principal weakness of the hypotheses that we have considered so far, is that they are unable to provide a satisfactory explanation for the great distances of the planets from the sun and their almost circular orbits round the sun. This difficulty can be removed if it is assumed that the materials which formed the planets were far removed from the sun from the very beginning, or in other words, if it is assumed that the planets have not been born from the sun. Accordingly, H.N. Russell suggested that there was another companion star of the sun and the two together formed a binary star or a twin-star system. A third star happened to pass close to the companion star of the sun, and this resulted in the ejection of gaseous matter from the latter in the form of a filament which ultimately separated from it.

The planets were formed from this gaseous filament of matter in course of time. In the beginning the planets were closer together and the satellites owe their birth to the mutual gravitational attraction between them. This third star was too far away from the sun to have any impact on the latter.

The suggestion that the primitive sun was a binary star cannot be dismissed as mere imagination, because at least 10 per cent of the stars in the universe are binary stars. In fact, in the opinion of some scholars, the number of binary stars is probably 30 per cent of the total. This hypothesis helps us to explain the great distances of the planets from the sun as well as their high angular momentum. But, the greatest weakness of the hypothesis is its inability to account for the removal of the sun's companion from its control and for the retention of the tidal filament which is later supposed to condense into planets and revolve round the sun. R.A. Lyttleton (1936) has supported this hypothesis and has tried to show by mathematical calculations that this is possible under certain circumstances. If it is assumed that: (a) the mass of the sun, the companion star as well as the third star was roughly equal; (b) the distance of the companion star from the sun was 1,700 million miles; (c) the speed of revolution of the companion star round the sun was 6 miles per second and the speed of the third star was at least 20 miles per second; and (d) the third star had come within a distance of approximately 3 to 4 million miles of the companion star. Under the above stated conditions, it is possible that tidal filaments will be ejected both from the companion star and the third star, the elliptic orbit of the companion star will change into hyperbole orbit and it will escape from the gravitational control of the sun.

Another objection raised against this hypothesis, is its failure to account for the present position of the planets, that is, their distance from the sun and their orbit, if it is assumed that the planets were formed at roughly equal distance from the sun as the hypothesis apparently seems to suggest.

The Nova Hypothesis of Hoyle and Lyttleton: In 1939, the British astronomer F. Hoyle put forward a new hypothesis according to which the planets of the solar system have originated from the explosion of a supernova. According to this hypothesis, the sun had another companion star which was much bigger than the sun, and the two together constituted a binary system. There was a terrific explosion in the companion star. Such explosions do happen in space. It is estimated that in our galaxy, i.e., the Milky Way, about twenty stars suddenly explode with brilliant light and become abnormally luminous. Such stars, with short abnormal luminosity, are called *nova,* and when the explosion and light are extremely intense, they are known as *supernova.* The explosion in the nova or supernova is possibly related to atomic reactions. When there is explosion, enormous amounts of gaseous matter are thrown out of the nova/supernova and its expansion becomes so great that suddenly its luminosity increases by hundreds of thousands. When the ejected gases become cold, the luminosity of the star decreases and it becomes less bright.

In the opinion of Hoyle and Lyttleton the planets of the solar system have originated from the expanding gases obtained from the explosion of a supernova. The expansion of the exploded gases was more in the direction of the sun as a result of which this gaseous matter came within the gravitational attraction of the sun, and later on it condensed to form the planets and the satellites in course of time. As a reaction to the explosion the supernova recoiled and started receding away from the sun and ultimately separated. This hypothesis is able to explain the high angular momentum of the planets, but is unable to offer any satisfactory explanation for the rotational movement of the planets as also for the origin of the satellites.

Interstellar Dust Hypothesis of Schmidt: In 1943, the Soviet scientist Otto Schmidt proposed a new hypothesis for the origin of the solar system, basing himself on the earlier views of Kant and Laplace. Large quantities of gases and dust particles are found scattered in the universe. Schmidt is of the opinion that in the beginning the sun was able to attract some gases and dust particles on account of its gravitation. This sheet of cloud of dust and gases started revolving round the sun. In the initial stages the cloud particles of different sizes were revolving round the sun in an irregular and unorganised manner. But, subsequently, the heavier particles collected near the bottom of the cloud heap and the cloud cover took the form of a vast flat saucer or plate. Due to mutual collision the particles started consolidating by condensation.

On consolidation they formed the embryos of the planets and later on took the form of asteroids. These asteroids started revolving round the sun within the dust disc in the direction of the latter.

The asteroids collected and absorbed the scattered matter by gravitational attraction which increased their size and they turned into planets. Even after the formation of the planets, sufficient material remained in space in attenuated (immature) condition which started revolving round the planets and in due course condensed to form the satellites.

Schimdt has taken the help of mathematics and logic in presenting his hypothesis. According to him: (i) The planets have originated not from the sun but from clouds of dust and gases. It is, therefore, not unnatural to find marked difference in the angular velocity of the sun and the planets. Further, the difference in the angular velocities of the planets themselves can be explained with reference to the differences in the angular velocities of the dust and gaseous particles at the time of their condensation, (ii) The inner planets of the solar system are composed of heavier elements (silica, iron, aluminium, etc.) while there is predominance of lighter elements (hydrogen, helium, methane, etc.) in the outer planets. According to Schimdt, when in the beginning the gas and dust particles collected round the sun in form of a saucer, it was not possible for the sun's rays to go far after penetrating the saucer.

In other words, the inner side of the saucer was hotter, and here the planets with heavier elements were formed. On the outer side of the saucer where the temperature was lower, the lighter gaseous materials condensed to form the planets on account of the cold, (iii) Schimdt has shown that the distances between the different planets is in accordance with statistical laws, because bodies of different size and speed will consolidate only at certain fixed distances in their revolution round the sun. (iv) The orbit of the planets of the solar system is circular. The reason for this is that the different particles after collision will attain their average motion. Having attained their average motion, the orbits of the planets would be similar and they will move in the same direction.

Although it is possible to solve a number of problems with the help of Schimdt's hypothesis, it has not been fully accepted. One major objection against this hypothesis is that it is not able to explain how the sun came to attract the clouds of gas and dust in the first instance.

Protoplanet Hypothesis of Kuiper: The American astronomer, Gerald Kuiper suggested certain amendments to the hypothesis of Weizsacker and in doing so put forward his own protoplanet hypothesis in 1951. In his opinion, the condensation in the nebulous ring round the primordial sun took place earlier than the condensation of the sun. In other words, the planets were created earlier and the present form of energy producing and light giving sun came to develop later. There was no difference in the composition of the nebula and the primordial sun. Both consisted primarily of hydrogen and helium with admixture of not more than 1 to 2 per cent of heavier elements. As the nebula contracted, it became less and less stellar and divided into a number of separate clouds of protoplanets. In the case of each original planet, the solid particles collected in the centre round which there existed an extensive envelope of gases. The protoplanets were of different sizes but they were much bigger than the present planets.

According to this hypothesis, as the protoplanets contracted the satellites were also formed close to the planets through a similar process. That is why this is the direction of revolution of all the planets. In the beginning the shape of the protoplanets was elongated on account of the tidal attraction of the sun, and their long axis kept pointed towards the sun. This made the direction of rotation the same as the direction of revolution. In other words, their periods of rotation and revolution were once equal.

In the meantime, the central mass was slowly condensing and developing into a star (the sun). Due to atomic reactions tremendous heat was generated in its interior parts and intense radiation started from the sun. When the sun became luminous, the solar wind of radiations and ejected particles drove away into space the remaining part of the nebula as well as the lighter gases of the protoplanets, and the remaining heavier parts of the protoplanets assumed the form of planets. The gaseous part of the protoplanets completely disappeared from the nearer planets, with the result that their mass and density became more. But only a part of the hydrogen-helium gases could be blown off the surface of Jupiter, Saturn and the outer planets. Thus, their density is relatively lower and especially the two largest planets—Jupiter and Saturn—have the lowest density. Gradually, the size and mass of the protoplanets decreased due to contraction and their speed of rotation increased. As a result of decrease in their mass, their gravitational attraction also decreased and their satellites moved further away from them.

We thus find that at the present time the basic assumptions of Kant and Laplace are once again being accepted. It is now believed that the origin of the solar system is not the result of some special accident or catastrophe but is due to the systematic contraction and condensation of a solar nebula. Since about 1943, there has been a swing back to theories of Laplacian type, in which the planets formed from a rotating gaseous nebula or dust cloud either around the sun and perhaps derived from it or else parental to the sun which formed during the condensation processes. The objection to Laplacian theories that the sun's angular momentum is too small has not been overcome. In the original gas cloud any viscous content in the gases would have slowed down the rotation at the interior of the nebula and accelerated the movement at the edge. Magnetic forces could have achieved the same result.

As opposed to Laplace, however, the original nebula is believed to be relatively cold. Famous physicist and Nobel-Laureate Harold C. Urey has supported the views of Kuiper on the basis of his own findings.

After discussing some of the important hypotheses, it may be concluded that the origin of the earth and the solar system is beyond the realm of certainty. Reliability, confidence and authenticity about any of the hypotheses and theories are unjustified. Much has to be learnt still before a definite theory about the subject is advanced.

However, a certainty had been expressed about the Laplacian hypothesis which dominated the field from 1796 till recent times. Its shortcomings, particularly its inability to explain the great distance of outer planets from the sun, led to the birth of other nebular and tidal theories.

The main drawback in the Laplacian hypothesis has, however, been removed. It related to the great distances to which the planets have been thrown from the centre of the nebula which the sun was. It is believed that such throwing away of the planets could be due to the magnetic force of the sun which caused ionisation of the particles of the envelop and increase in their centrifugal force.

Weizsacker's dust cloud hypothesis, based upon new knowledge about the chemical composition of cosmic matter, is a modern version of the Laplacian hypothesis which, therefore, stands revived. The Laplacian hypothesis still seems to be more reasonable about the origin of the earth as well as the solar system.

Primitive Theories

Mankind has long taken interest in the nature and origin of the rocks. Among the ancient Greeks, Herodotus (485-425 BC) recorded the presence of marine shells far inland in Egypt and concluded that they had been left there by the retreating waters of some earlier sea. Pythagoras (6th century BC) is said to have taken the presence of fossil marine creatures high in the mountains of Greece as evidence of the elevation of a former sea bed, and Strabo (63 BC-AD 23) certainly accepted such a view. As the Mediterranean Sea lies in one of the world's volcanic and seismic belts, many Greek and Roman authors were fascinated by the phenomena of volcanoes and earthquakes. Strabo, in common with many later writers, saw volcanoes as natural safety valves designed to permit the escape of dangerous terrestrial vapours. Likewise, Seneca (AD 3-65) regarded earthquakes as the result of some vapours becoming turbulent within the earth's supposedly cavernous interior. In China, in 132 AD, Change Heng (78-139 AD) invented an 'earthquake weatherock', the first seismograph ever constructed. Rather later, Ibn-Sina (Avicenna 980-1037 AD), the father of the earth sciences in the Arab world, discussed not only earthquakes but also a wide variety of other geological phenomena. He also displayed a sound understanding of many earth processes at a time when, in western Europe, Christian scholars were more interested in saints and sermons than in sandstone and sapphire. Some of the ideas then current in western Europe must now strike as bizarre and ridiculous.

To some extent Europeans were drawn to earth sciences in the aftermath of the Renaissance because of the obvious application of those sciences during an age in which the demand for minerals increased.

It was during the 17th century when scholars began to formulate theories about the origin of the earth. These were attempts to devise convincing and fairly detailed histories of the earth. The French philosopher Rene Descartes (1596-1650) was an early influence in this new genre, but geologically the most important of the theories was that published in Florence in 1669 by Nicolaus Steno (1638-86), a Dane resident in Tuscany. Steno illustrated his theory by means of a series of diagrams, representing successive stages in the evolution of the earth. He displayed an understanding of many important geological concepts.

The theories devised by the 17th century cosmogonists continued to excite much attention during the following century, but slowly it came to be appreciated that such theories were nothing more than imaginative exercises.

The Concept of Nebula: Nebula (Latin 'mist') is a term formerly used in astronomy to denote any object situated beyond the solar system that is seen as a bright or dark area in contrast to the point-like images of stars.

Nebula differ greatly in size. The only one easily visible with the naked eye is the beautiful *Orion Nebula.* It is the brightest and most thoroughly studied diffuse nebula. Some nebulae are extremely large; the largest, the *Gum Nebula* (named after its discoverer, the Australian astronomer Colin S. Gum) measures 40° in angular diameter.

Naked-eye observations of 'cloudy stars' were mentioned by the Greek astronomers Hipparchus (190-125 BC) and Ptolemy (AD 138) but these objects are known to have been compact clusters of stars. In 1610, two years after the invention of the telescope, the *Orion Nebula,* which looks like a star with the naked eye, was discovered by the French scholar and naturalist Nicolas-Claude Fabri de Peiresc.

The 20th century has witnessed enormous advances in observational techniques. Bernhard Schmidt invented an extremely fast, wide angle camera that made possible the detection of large faint nebulae. The use of space satellites has permitted the study of ultraviolet and x-ray portions of the spectrum that would otherwise be absorbed by the earth's atmosphere.

Modern Theories of the Origin of the Earth and the Solar System: The theories about the origin of the earth and the solar system have generally been classified under two categories, i.e., (i) the evolutionary or natural or monistic hypotheses which suppose that the system of planets has evolved from stars, and (ii) the cataclysmic or catastrophic dualistic which believe in some sudden and violent event in the space like the collision or close approach of two stars.

Laws of Evolution

The Nebular Hypothesis of Kant: The famous German philosopher, Immanuel Kant (1724-1804), who anonymously published his views in 1755, was the real propounder of the nebular hypothesis. His hypothesis about the origin of the earth was based on Newtonian Law of Gravitation. He believed that the hard particles of super naturally created matter collided with one another by gravitational attraction, and generated heat and rotation in this process. In this way, the original static and cold matter was converted into a nebula (vast hot gaseous mass) rotating with such great rapidity that strong centrifugal force was created about the equatorial plane. This led to the throwing of successive rings of matter. The rings condensed into planets. The planets underwent similar spinning and threw off rings which became their satellites. Nine successive rings came out of the nebula in this manner and they ultimately formed nine planets. What remained of the original nebula became the sun. Thus, according to Kant, the earth has originated by the aggregation and solidification of the materials which separated from the nebula in the form of a ring on account of centrifugal force. Through repetition of the same process, circular rings also separated the planets resulting in the formation of satellites. It is in this manner that, in the opinion of Kant, the entire solar system originated and got organised in the existing form.

The theory of Kant has been criticised on several grounds. Some of the important criticisms of Kant's hypothesis are as under:

1. Kant attempted unsuccessfully to explain the rotation of the nebula while his theoretical successor', Laplace, assumed that the nebula was already in a rotating state.
2. Kant's hypothesis has been found to be against the mathematical laws. Kant's assumption that the mutual

collision of particles resulted in the generation of rotational movement in the nebula is against the principle of conservation of angular momentum.

3. The second assumption of Kant that the rotation of the nebula increased in speed with increase in the size also goes against the above law, because the speed decreases with the increase in size, and when the speed increases, size will decrease.
4. Kant did not explain the source of the origin of the primordial (fundamental or primary) matter.
5. Kant also did not explain the source of energy to cause random motion of the particles of the original matter which was cold and motionless in the initial stage. According to Newton's first law of motion, "a body remains at rest, or if in motion it remains in uniform motion with constant speed, unless and until an external force is applied on it". The particles of the primordial matter, as assumed by Kant, were at rest and no external force was applied on them, then what was the cause for the random motion among the particles of the primordial matter. Despite all these shortcomings, the hypothesis of Kant is highly commendable. He put forward his theory without the knowledge of the gases, and the elements, which have been gathered during the last about two-and-a-half centuries (1755 AD). The significance of his hypothesis lies in the fact that his was the first hypothesis based on Newton's law of gravitation and thus he was the pioneer of the nebular hypothesis. It has, therefore, been commented that his hypothesis was a sort of *'lucky guess'*.

The Nebular Hypothesis of Laplace: Marquis de Laplace put forward his nebular hypothesis in 1796. This theory has had its dominance till recent times owing to its apparent rational appeal. His views had great resemblance with those of Kant, but in all probability he had no knowledge about Kant's publication.

Laplace started with the assumption that the rotating nebula which gave birth to the solar system was already in existence. The main points of the theory are as under:

1. There was a hot rotating nebula in the space of the universe. The hot rotating nebula under went cooling and contraction,

and consequently its speed of rotation increased. As the speed of cooling and contracting nebula increased, the centrifugal force of the outer rim in its equatorial region increased and it got separated from the main body and something like the present ring of Saturn developed. This ring, later, is believed to have collected as a planet. Further cooling of the rotating nebula would follow. Consequently, it would contract, and its speed of rotation would increase. The centrifugal force would go on increasing and the force of gravity on the equatorial outer rim would be overcome by the centrifugal force. The rim would separate and form another planet, the continuation of this process is believed to have caused the formation of the different satellites.

2. The repetition of the same process on the different planets is believed to have caused the formation of their moons or satellites. What remained at the nucleus of the nebula became the sun.
3. The above given process caused the rotation of the planets, and their revolution in the same sense and their location roughly in the same plane. No doubt, some of the satellites have motion opposed to that of most of the planets and their moons, and this fact goes against the hypothesis.
4. The evolution of a solid earth from an originally gaseous and later liquid mass is apparently supported by a solid crust on the upper part of the earth and the evidences of a liquid molten interior (e.g., vulcanicity).

The hypothesis of Laplace is simple and more logical. With its help, it was possible to explain the uniformity in the direction of rotation and revolution of all the planets and satellites (with the exception of a few satellites). This hypothesis also explains why the orbits of the different planets were roughly circular and in the same plane. According to this hypothesis, the earth and other planets were first in a gaseous state, and later on they passed through a molten stage and finally the solid crust was formed. The increase of temperature in the interior of the earth as well as vulcanicity support this hypothesis.

For all these reasons, the hypothesis of Laplace is being appreciated now for more than 200 years. The nebular hypothesis as advocated by Laplace has also been criticised on several counts. Some of the main criticisms are as under:

1. It is more likely that the separation of matter from the original nebula was a continuous phenomenon. The particles of the gaseous nebula were loose and had little cohesion among themselves. Therefore, instead of breaking away intermittently, they might more probably have developed into one mass and then separated. The cooling and contraction of the original nebula would probably have developed into two stars and not into a solar system like ours.
2. He did not describe the source of the original nebula.
3. Laplace did not discuss how a separated ring would aggregate into one globular mass, i.e., a planet. A more probable development appears to have been a number of more or less equal smaller planets in the same orbit.
4. The hypothesis of Laplace does not explain the direction of movement of the satellites of the planets, Uranus and Neptune, which is opposite to the direction of motion of the remaining bodies of the solar system.
5. He could not explain why only nine planets came into existence.
6. Laplace could not explain adequately the force that might cause the separation of matter from the original nebula. The angular momentum (which is proportionate to the mass of a body, its velocity of motion and its radius of revolution) of parts must be equal to the angular momentum (i.e., the force of motion of a body as it goes round a point) of the whole.
7. When the planets started forming, the amount of matter that separated from the sun must have been a very small fraction in the case of each planet. This is clear from the present size of the planets. The original angular momentum of each of the planets must have consequently been a very small fraction of that of the sun. Thus, the concentration of 98 per cent of the angular momentum in the planets cannot be comprehended in this way also, and thus the nebular hypothesis was rejected.

8. If the sun is the remaining nucleus as claimed by Laplace, it should have a small bulge around its middle part (equator) which would point out the probable separation of irregular rings from the sun, but there is no such bulge in the middle part of the sun.
9. James Clark Maxwell, the British physicist, proved with the help of his investigations that if we assume the total mass of all the planets to be a ring around the sun, then this mass is so small that it was impossible for it to condense into planets under the influence of gravitational attraction.
10. The hypothesis of Laplace does not explain how the gaseous rings collected into rounded planets. According to the law of dynamics, it is more probable that the rings would break into smaller and roughly equal parts and thus several small planets would be formed in the same orbit. Despite all these omissions and commissions the nebular hypothesis of Laplace explains many of the difficult questions about the origin of the solar system. It is because of these points that this simple hypothesis is still considered as one of the most reasonable conjectures about the origin of the earth and the solar system.

Revival of the Laplacian Nebular Hypothesis: The Laplacian hypothesis has been appreciated after 1943 because of the discovery of some new facts. Since then there has been a revival of this theory and a swing back to this and similar other theories. The reason is that the objection regarding the angular momentum (or the great distances separating the planets from the sun) has been overcome. The reasons are as follows:

1. In the original gas-cloud or nebula, the drag of any viscous matter would have retarded the rotation in the inner part of the nebula and would have increased movement at the outer margin. It might have resulted in greater centrifugal force in the outer planets and their throwing away to great distances.
2. Such throwing away of the planets to long distances could have also been achieved by magnetic force. It is thought that the magnetic field of the original sun was considerable. The cloud of gas around it would have been mostly of hydrogen. Some of this hydrogen might have been ionised (i.e., charged

electrically) and the hydrogen ions could act as magnets. If a large rotating magnetic body is surrounded by small magnetic particles, it may carry the latter along with it. If the sun rotated rapidly within a period of few hours, it would drag the magnetised hydrogen ions of the surrounding gases rapidly. This would mean the increase of centrifugal force and throwing out of the finer matter of the dust-cloud envelop of the sun. The inner region now occupied by the inner planets of the solar system would have been left with larger solid masses.

3. The large part of the momentum of the sun was transferred to the gas ions. This slowed down the rate of rotation of the sun. Thus, the original matter of the solar system was divided into two zones: (i) an outer zone rich in gas, and (ii) an inner zone without gas but with solid particles. The inner zone became the region of denser inner planets and the outer region became that of lighter planets.

There is, however, uncertainty as to the manner of growth of planets from clouds of gas and dust. There is relative dearth of inert gases like argon, helium, neon, xenon and krypton on the earth. This suggests that such gases did not come from the earth and that the formation of the inner planets was from solid matter at low temperature conditions.

The atmosphere and the oceans of the earth are derived from its interior. Water came to the original earth not as a gas but as solid hydrated minerals like mica.

Similarly, there is uncertainty about the temperature and pressure conditions under which the planets were formed. It is guessed that the earth aggregated as a combination of metallic iron silicates, particularly magnesium silicate and iron sulphides.

Cooling Antiquity

Although different authorities vary in their presentations of the cooling history of the earth, they mostly agree that the beginning was marked by 'cool' conditions and the sequel was a long period of heating till the present temperatures were reached by the earth. After the attainment of the present temperatures thermal uniformity has been maintained.

Inference from Meteorites: It is believed that the meteorites originated from the disintegration of asteroids (about a thousand minor planets between Jupiter and Mars). These meteorites may represent the sample of matter of which the universe might have been made.

When these asteroids, the source of meteorites, were formed, the temperature of the original matter is believed to have been definitely low. Then, there was extremely rapid radioactive heating of the parental asteroid material. Subsequently, these bodies cooled rapidly. The radioactive parental material does not exist now. But some decay products can be identified.

The occurrence of such short-lived radioactive species suggests *supernova-like* explosion or similar nuclear event before the origin of the planets. If the earth is believed to have developed on the pattern of meteorites, then our planet would have become radio actively heated at an early stage. Subsequently, it acquired a steady thermal condition in which radioactive heating was practically balanced by the loss or flow of heat from the surface of the earth.

Landforms after Ice

Landforms that are Periglacial

The result of various processes is the creation of several landforms, which can be divided into two parts, major and minor landforms.

Major Landforms

Block Fields: Blockfields, known as felsenemer (literally rock sea) in German, are produced by block disintegration generally by frost action (congelifraction). These are continuous spreads of broken angular rock fragments of boulder dimensions which mantle the high mountain or plateau. On steep slopes, beyond a critical angle, the blocks will tend to move down the mountainside.

Rock Streams or Rock Glaciers: Rock streams or rock glaciers are transitional landforms between periglacial and glacial regions. They are accumulations of unsorted, till-like coarse to fine debris, with an ice core and or interstitial ice having a glacier shape and spreading down valley. The debris moving down valley in the shape of a glacier assume lobate, tongue like or spatulate forms.

Rock glaciers are composed principally of material provided by rock slides, rock falls and avalanches and also by glacial, fluvial and aeolian processes to some extent. Rock glaciers can be as long as 3 km and can have a height of 60 m at the toe.

The surface of the rock glaciers is characterised by linear furrows and intervening ridges, scattered depressions and patches of fresh and permanent snow. Fairly thick, large angular debris abound in the surface material of rock glaciers. The frozen material with large content of ice in pore spaces deforms plastically as it moves. Rock glaciers move very much like glaciers as they move forward by internal deformation and debris slide of their termini, resulting in a rolling or continuous-tread form of transport. The important point of difference between the rock glacier and glacier is that the former is mostly rock debris with interstitial ice where as the glacier is mostly ice with included rock fragments.

Stone Streams: A stone stream is an accumulation of angular fragments and glacially driven boulders usually in valleys and only occasionally beyond valley divides. Although all stone streams originate in periglacial regions not all of them are of periglacial origin. Abundant angular fragments are derived from frost susceptible rocks. Fine grained particles can be washed out from the rock rubble. The debris in stone streams are well sorted, the upper portion dominated by large debris while the lower portion is dominated by small debris. This sorting is due to the process of frost heaving. Stone streams move downslope by the force of gravity, frost heaving and solifluction. In many cases stone streams appear to be downward extension of blockfields occupying summit areas.

Asymmetric Valleys: Asymmetric valley, as the name suggests, is a valley in which the slope on one of the sides is greater than that on the other. It is formed by differential weathering of slopes under periglacial conditions. This occurs when valley alignments lead to differing degrees of exposure to direct Sunshine and shadow and the processes operate with different efficiencies on the Sunny and shaded slopes. But there is no complete agreement about the precise nature of the controls. Besides insolation, the prevailing wind direction (as it affects snow drifting) is equally important. Asymmetry can be produced in many ways one side may be steepened more than the other, one side may be flattened more than the other or one side may be steepened while the other is flattened. In general, the south facing slopes in the Northern Hemisphere are more steeper than the north facing slopes.

On slopes facing south and southwest, greater insolation is likely to produce greater temperature fluctuations, more freeze-thaw cycles and therefore, a much more active debris mantle leading to steepening of the slope length. At the same time gelifluction is more intense on the valley side facing the Sun for the longest time each day. Thus, shaded side stays colder, less prone to gelifluction, and the slopes are steeper. Also, the copious colluvium from the south facing slopes forces river to preferentially erode and steepen their north facing valley walls. It has also been suggested that the gentler slopes facing northeast are due to shelter from westerly winds permitting a thicker snow cover to accumulate. The slow release of meltwater would then provide abundant moisture to promote the downhill movement of the debris mantle. However, steepening can also be on the southern slopes as is evident from the cross sections in Alaskan region. All in all, the valley side systems are more complex and there is need of further research in the periglacial environment.

Cryoplanation Terraces: Cryoplanation terraces, altiplanation terraces or goletz terraces (in Russian) are irregular features found in highland regions ranging upto 1000 m in length, several hundred metres in width and 5 or 10 m high at the frontal edge. Many of the terraces are cut into solid rock and are formed due to nivation, acting around snowbanks, settling in existing hollows and on steps. Solifluction may help to remove the debris formed by nivation.

Medium and Minor Features

Ice Wedges: Ice wedges are narrow cracks or fissures in the ground infilled with ice which may extend below the permafrost level. These are V-shaped, having a typical vertical orientation and a milky white appearance (because of the presence of many small air bubbles). Ice wedges originate due to thermally induced cracking of permafrost on an annual basis. Intense winter cold (-15°C) leads to contraction in the upper permafrost horizon creating minute fissures. During winter, hoarfrost accumulates in the cracks while during spring, water from the active layer percolates down to freeze below the permafrost table forming ice veins. As the temperature of the permafrost rises, the thermal expansion causes the sediment to deform and upturn against the ice veins. Renewed thermal contraction in the next winter reopens, the crack and the whole process is repeated.

Each year the crack reopens, fills and freezes, creating a wedge of ice called ice wedge. Ice wedge networks are best developed on horizontal surfaces composed of fine alluvial sediments. They are, therefore, particularly conspicuous on river terraces and flood plains. They can be either epigenetic or sygenetic type. Epigenetic types are those that have developed after fluvial aggradation has ceased, while syngenetic types grow concurrently with sedimentation and extend upwards as the level of ground surface is raised. These can grow to a width of 60 cm or more and depths of 1 to 5 m and more.

Ice wedges grow slowly, viz., < 0.25 cm annual increment of ice in ice wedges. Their growth is arrested when sufficiently thick gelifluction debris and snow insulate the surface from seasonal temperature changes. The winter stress, in relation to the strength of ice wedges, remains low due to which a crack alongside the existing wedge fails to open up. The wedges also cease to grow when the stress is relieved by the development of ice wedges adjacent to each other. If there is an amelioration in climate, the ice infilling may disappear and the ice wedge cast become infilled by sediments so that its former location is seen in section. It is known as fossil ice wedge or ice wedge pseudomorphs.

Aerial photographs have disclosed very extensive ice wedge networks. In Russia, ice wedges in some regions are 60 m deep and extremely large at the surface. In North America, the feature tends to be small, 1 m or more at the top and about 6 m long in the permafrost. Ice wedges form large scale polygonal patterns when they reach the surface.

Sand Wedges: Not all wedge-shaped forms originate in an ice wedge manner. The decay of large tree roots can produce features resembling fossil wedges. Sand wedges are one such feature. In contrast to ice wedges, which represent the humid polar climates, sand wedges suggest an environment of comparative aridity and strong winds. These develop by continual infilling of frost cracks with wind blown sand. Each year more sand is added to the growing structure until eventually it may reach a width of over 1 m. These sand wedges are widely found in Antarctica.

Ice wedges form polygons on a large scale but polygons are also formed on intermediate and small scales.

Involutions: Involution refers to the deformation of stratified deposits of unconsolidated materials just below the ground surface in permafrost areas. Sometimes deformation is so pronounced that the original structure becomes indistinct and untraceable. These structures are known to occur in alternating beds of variable grain sizes such as silt, clay and sand. In plan, many such structures are circular or polygonal. Involutions do not readily develop in beds of uniform grain size. The formation of involutions is caused by a number of factors and the different explanations for their formation include—

Intergranular pressure changes during thawing.

1. Expulsion of pore water to unfrozen sediments during freezing.
2. Differential volume changes due to freezing of pore water. If the adjacent material remains unfrozen, it is squeezed and eventually engulfed by the expanding frost-heaved material.

Patterned Ground: Patterned ground is a group term for more or less symmetrical forms such as circles, polygons, nets, steps and stripes that are characteristic but not necessarily confined to permafrost regions. For its optimum development, patterned ground requires a combination of moderate moisture and frequent cycles of freeze-thaw action but it also depends on the susceptibility of the soil to frost sorting and gelifluctions and the presence or absence of vegetation. A patterned ground is polygenetic. Based on geometric forms there are five major types of patterned ground form.

1. Circles occur in isolation or in close groups. In circles, typically a circle of coarser stones surround a centre of fine materials. They can be up to 3 m in diameter and the dimensions of sorted and non-sorted circles are almost similar.
2. Nets are smaller in size. They are actually concentrations of circles which have not reached the stage of interfering with each other. They can be both sorted and unsorted and

represent a transitional stage between true circles and true polygons.

(c) Polygons are the best known type of patterned ground. They occur as part of a mesh of sorted and non-sorted materials on nearby horizontal surfaces. Sorted polygons have a border of stones surrounding an area of finer material; while in non-sorted polygons the border of stones is lacking. Their borders may be raised above a low centre but more commonly they occupy a perimeter furrow around a slightly domed centre of finer material with the topographic difference, often picked out by differences in vegetation.

Characteristics of Different forms of Patterned Ground

Circles	Occur singly or in groups. Typical dimensions 0.5-3m. Non-sorted type characteristically rimmed by vegetation, sorted type bordered by stones which tend to increase in size with size of circle. Found in both polar and alpine environments but are not restricted to areas of permafrost. Unsorted circles also recorded from non-periglacial environments.
Polygons	Occur in groups. Non-sorted polygons range from small features (< 1m across) to much larger forms up to 100 m or more in diameter. Sorted polygons attain maximum dimensions of only 10 m. Stones delimit polygon border and surround finer material. Non-sorted forms are delineated by furrows or cracks. Some types of polygon occur in hot desert environments but most are best developed in areas subject to frost. Ice-wedge polygons only form in the presence of permafrost.
Nets	A transitional form between circles and polygons. Usually fairly small (<2 m across). Earth hummocks comprising a core of mineral soil surmounted by vegetation are a common form of unsorted net.

Steps	Found on relatively steep slopes. They develop either parallel to slope countours or become elongated downslope into lobate forms. In unsorted forms the rise of the step is well vegetated and the tread is bare. In the unsorted type the step is bordered by larger stones. Lobate forms are known as stone garlands. Neither type confined to permafrost environments.
Stripes	Tend to form on steeper slopes than steps. Sorted stripes composed of alternating stripes of coarse and fine material elongated down slope. Non-sorted variety delineated by vegetation in slight troughs. Not confined to periglacial environments.

Source: Based on discussion in A.L. Washburn (1979 1 Geocryology. Edward Arnold, London, pp. 122-56.

These polygons can be of three types—Ice wedge polygons, Stone polygons and Breckland Polygons.

1. Ice wedge polygons cover many thousands of square kilometres. Each polygon, from 3-30 m in diameter or more, is outlined by shallow trough beneath each of which is a wedge shaped foliated mass of ice, a metre or more broad on the top and extending downwards from 3-7 m or more. Their formation is the same as that of ice wedges.
2. Frost crack polygons are less conspicuous and are outlined by open cracks or shallow depressions but not underlain by ice wedges. Frost crack polygons are common on mountainous slopes with intense freeze-thaw activity. In seasonally cold regions, thermal contraction of the ground causes cracks to develop in intersecting directions forming tetragonal, pentagonal and hexagonal forms. Frost crack polygons represent well-sorted debris formed under conditions of repeated freeze-thaw activity or multi-gelation. The annual freeze-thaw cycle produces large lateral stress in the active layer. The stress causes the various rock fragments to be pushed upward and outward to the edges of the freezing

surface which operates from above. This repetition causes the stones to rise to the surface. Fine particles force their way into the voids between the stones forming a sorted pattern.

3. Breckland polygons are of unsorted type, intermediate in size between stone and ice wedge polygons. The average distance between polygon centres is 10.5 m and between the axes of stripes which replace polygone on slopes exceeding 1-2 degrees is 7.5 m. These polygons are bounded by shallow troughs and are formed primarily from an uneven distribution of sandy material. Their origin, however, is not clearly understood.
4. Steps occur on gentle slopes where polygons become elongated into small terrace like features. Their occurrence is not always necessary and polygons may change directly into stripes. In the sorted type of terrace, a bank of coarser stones forms the front of the terrace.
5. Stripes are elongations of polygons and steps on slightly steeper slopes, i.e., coarse and fine sediment trend downhill. The excellent drainage provided by stone stripes enables hillslopes to remain stable at unusually steep angles perhaps to 30° without slumping or sliding.

Hill of Ice

Ice Mounds are known by various names. The larger ones are called pingos (an Eskimo word in Arctic, North America and Greenland), bulgunnjakh (Russian) or hydrolaccolith, while small forms usually arising in peat bogs have been called palsas.

1. *Pingos:* The term pingo is used by Eskimos of the Mckenzie delta in Canada to denote scattered dome shaped hills rising sharply above the alluvial plain. Pingos range from minor hillocks only 2 m high to over 50 m high. Diameters vary from only 10 m to over 200 m, but the sides are almost always steep and often exceed 20° in angle. The smaller ones are almost perfect dome shaped, but the larger ones are breached by crater-like depressions 5 m or more in diameter.

   On the basis of their association with the nature of permafrost and the mode of their formation, pingos can be classified as closed system or Mckenzie type and open system or East Greenland type.

a. Closed system or Mckenzie type pingos are associated with the sites of former lakes. Similar features in northern Siberia occur in the same type of situation. Closed system pingos develop when permafrost advances into a shallow lake basin. The lake prevents permafrost from forming below it for some time, but ultimately filling and shallowing of the lake will allow permafrost to develop on its floor. At this stage, the permafrost at the sides of the lake starts to extend inwards. The saturated unfrozen sediment is trapped between the downward freezing surface from the lake bottom and the advancing permafrost from the sides of the lake. The continued saturated mass of sediment gradually freezes. Excess pore water is expelled by the cryostatic pressure exerted by the advancing permafrost. At the same time, water freezes as ice lenses, ice core and segregated ice beneath the frozen ground. The pressure caused by the permafrost advances and the growth of ice crystals updome the sediment. To relieve the pressure the overlying layers are forced up and finally a large mass of ice freezes in the core of the dome shaped hill that has evolved. This type of pingo has been referred to as closed system pingo because it is formed from an enclosed unfrozen pocket trapped by advancing permafrost.
b. Open system or East Greenland type pingo is virtually an artesian feature. These pingos occur in thin discontinuous or sporadic permafrost and on sloping surfaces and are a common occurrence in unglaciated parts of Alaska. Surface water penetrates through thin permafrost to reach the talik. At some low point on the aquifer, artesian pressure builds up sufficiently to force the circulating groundwater to move vertically towards the surface where it refreezes as injection ice eventually forming a core of clear ice. The result is the updoming of the surface over it forming a pingo. Since the artesian pressure can be released through more than one outlet, several small-sized pingos can form in a given area.

That pingos can form in this way is less likely because artesian pressure is never enough to create pressure leading eventually to doming. Moreover, pockets of water beneath the pingo ice in artesian tubes have not been found. Therefore, the cause of open system or East Greenland type pingos is not adequately understood.

There is considerable diversity in the rates at which pingos grow. Some of the very small examples on the Mckenzie delta are known to have formed since 1950 and growth rates as high as 0.5 m per yr. have been recorded, while the larger pingos grow not more than 0.5 mm per year.

When a pingo grows to a certain size or when the pressure from beneath exceeds a certain critical value, the surface cracks and exposes the ice core. This eventually, leads to melting and subsidence of the top of the dome into a crater. When the crater finally decays it leaves a shallow pool surrounded by a rampart composed of sediment that was either pushed or slumped to the margin of the ice core.

c. Fossil pingos or pingo casts consist of a circular rampart probably breached in one place and encloses a marsh or pond. They are found in Asia, Europe and North America and are indicative of former permafrost conditions.

2. Palsas are mounds or ridges of peat containing perennial ice lenses and a core of permafrost. They are found mostly in discontinuous permafrost. They can be 10 m high, 150 m long and 30 m wide. They are found in Canada, Finland, and the adjacent territory of the erstwhile USSR, Iceland, Norway, and Sweden. The surface of palsas is usually criss-crossed with open fissures caused by frost cracking, dilation cracking (caused by updoming) or desiccation. The factors on which the development of palsas depend are associated with thermal properties of the organic terrain, and the subarctic environment experiencing long severe winters with little snow cover. The thermal conductivity of moist peat approaches that of ice and this favours the development of ice lenses. The development of epigenetic ice leads to frost

heaving and develop the form above the water table in the bog. Palsas differ from pingo in the sense that it has a characteristic presence of peat which is comparatively rare in pingos. Palsas begin to decay when the wastage of peat along the expanding dilation cracks, caused by differential heaving. It exposes the segregated ice and eventually merges with the general level of the bog without leaving any trace of the former feature.

Hummocks: Hummocks are mound of Earth on the surface of the permafrost having either a core of mineral soil (Earth hummock) or a core of stones (turf hummock) up to 70 cm in height and up to 3 m in diameter. They originate either due to (i) squeezing of the ground surface caused by lateral pressure exerted by freezing of active layer resulting in the formation of small knots, or (ii) due to frost heaving.

Thermokarst: Thermokarsts refer to the ground surface depressions which are created by the thawing of ground ice in periglacial zone. The subsidence results in the creation of some features which may resemble true karst features but only thermal processes are involved, there is no chemical weathering involved and it does not depend on the presence of limestone. The uppermost permafrost horizon is rich in moisture in the form of sub-horizontal ice lenses. If the permafrost table is somehow lowered, liquid water is released and the ground surface sags. This melting can be brought about as a result of human disturbance (deforestation, vehicle tracks, house construction), periodic surface flooding, an unusually warm summer, or climatic change. This results in the creation of depressions of irregular size and shape called thermokarst. The features include-thaw lake, alas and beaded drainage.

Thaw lakes originate when an ice lens is exposed to the air and melts. The lake quickly assumes an approximately circular form as it melts and undercuts the adjacent ice rich permafrost. There is a cycle whereby vegetation growth around the lake eventually protects the underlying ice from melting and the lake begins to be infilled and it revegetates. Eventually the dried out lakes may be marked out by the site of a pingo.

Alas is a large depression and is characterised by a flat, sometimes lake-covered floor and steep surrounding walls. It is well developed in Siberia and is a product of long term localised melting of the permafrost. The depressions may coalesce to form irregular linear troughs known as alas valleys, tens of km in length. Beaded drainage is characterised by interconnected pools and short linking streams, having the overall appearance of a necklace of beads. Pools may form by melting of ice at the junction of ice wedge polygons and streams follow aligned ice wedge patterns.

String Bogs: String bogs or strangmoore are areas of muskeg that consists of ridges (less than 2 m high) of peat and vegetation interspersed with depressions often filled with shallow ponds all occurring on terrain with gradient of less than 2°. These are quite widespread in the forest tundra mosaic. There is no clear agreement as to how they form but they may be formed in any of the following ways.

1. Solifluction and wrinkling of the bog surface,
2. Hydrostatic pressure rupturing the bog surface and frost heaving raising ridges,
3. Differential thawing of permafrost leading to local bog collapse and a combination of biological and hydrological factors.

Landforms, which are Glacial

The Meaning: When the temperature falls below 0°C, water vapour condenses and freezes into ice crystals which are precipitated in the form of snow. This snow is light, fluffy and easily blown away. High latitudes and high altitude regions receive snow in the winter, which gets melted in the subsequent summer. All the snow may not melt and some of it may accumulate, so that there is a perpetual snow cover. This happens in many parts including Greenland, Arctic, Antarctica and on top of some high mountains (Table). The level above which this snow cover is perpetual, is called the snow line. This snow line varies from a height of 4800 metres at the Equator to sea level around the Poles.

Since, atmospheric moisture decreases with altitude, the conditions may become unfavourable for snow formation at a particular height. Thus, an imaginary snow line can be drawn which will demarcate the regions that are snow free. The region between permanent snow line and imaginary line is called Hinosphere.

Ice on Earth

Ice Body	*Surface (km²)*	*Mean Thickness (km)*	*Volume (x10⁶ km²)*
Antarctic ice sheet	13.3.106	1.8	24.0
Greenland ice sheet	1.7.106	1.6	2.7
Other glaciers	-	-	1.3
Total ice on Earth	15.0.106	1.8	28.0
Additional amount of ice on Earth during last ice age	25.8	1.9	49.0

The Establishment

Glaciers are an open system like rivers. Water enters the system primarily in the upper part of the glacier where snow accumulates and is transformed into ice. The glacier can be divided into an upper accumulation zone and a lower ablation zone or wastage zone (the combined loss by evaporation and melting is called ablation). These two zones are divided by a firn line. Below this line, no firn is produced. In the glacier system, the accumulation of ice occurs at all points—on the tributary and the main glacier but it is likely to be the greatest at the head of the system. At the same time, the loss of ice by melting or evaporation (ablation) takes place over the entire system but is inevitably maximum in lower portion where temperatures are high. A particular form of wastage, occurs if glacier terminates in water. Here, great blocks of ice can break off and float away in a process called calving. This is the process that produces icebergs of the polar areas.

Ice masses will advance or retreat according to the balance between ablation and accumulation. At the head of the system, annual accumulation is greater than annual ablation. In the lower reaches, the situation is reversed. At the firn line, ablation and accumulation are equal. Ice moves continuously across the firn line to balance the volumes of ice on the two sides of the line. The snout or terminus occurs at a point where ablation is equal to the forward motion of the ice. Retreat of an ice front may be due to reduced precipitation or increased temperatures or both. It will be accompanied by thinning ice and extensive meltwater activity at the margin. Advancing ice front results from increased precipitation or cooling of atmospheric temperatures or both. It has to be borne in mind that the glacier may be said to, retreat or recede, but this is only a matter of speaking. It is not that ice moves backwards but the position of the terminus or snout.

System of Transformation

As snow accumulates, it is gradually transformed into ice. This transformation is the result of several processes, not all of which are clearly understood. Among the processes responsible for transforming snow into ice are—sublimation, recrystallisation (with growth of large crystals at the expense of smaller ones), (melting-refreezing under pressure) regelation and compaction under weight. The process of changing snow into ice is related to local conditions of temperature and pressure, for example, (i) a low annual average temperature, (ii) continuous and abundant snowfall, (iii) altitude of mountain region and latitude of locality, and (iv) existence of suitable landscape and relief features.

As snow piles up by actual precipitation and drifting, compaction of the lower layers takes place. The increasing pressure then causes a slight lowering of the melting point. This melted water percolates through the interstices of snow mass and refreezes (called regelation) around snow crystals to form ice granules. This process takes place more rapidly in temperate regions, where the temperature is close to 0°C and where there are longer periods of temperature above freezing point than in colder conditions of Arctic and Antarctica. When the permeability of fresh snow is reduced, the air pores are no longer connected, the mass is known as firn (or nevee) and there is an accompanying increase in density from about 0.1 g/ cu cm to about 0.5 g/ cu cm. Firn is dull white, impermeable and structureless form of ice (0.8-0.9 g cu km). As the pressure increases more with the overlying weight, the firn is converted into ice. This has very few air pockets and a blue sheen.

Ice is a very different substance from snow. It is plastic and can be deformed under pressure. A thick mass of ice resting on a hillside is under considerable internal pressure due to its own weight while at the same time gravity favours downhill deformation. Plastic flow results and the ice begins to move downslope. When ice thickness reaches 30 to 60 metres or more lateral movement occurs. This moving river of ice is a glacier.

The Categorisation

Glaciers are classified according to their morphology and internal temperature regime.

Morphologically, there are two types of glaciers—continental and valley glaciers.

Morphological Classification of Glaciers

Basic Type	*Component or Sub type*	*Characteristics*
Ice sheet and ice cap (unconstrained by topography)	Ice dome	A dome-like mass with a convex cross-profile formed in response to the basic flow characteristics of ice
	Outlet glacier	Glaciers which radiate out from an ice dome often occupying significant depressions. Within the ice dome, they can be distinguished by a zone of rapidly moving ice termed an ice stream
Ice shelf		A floating ice cap or part of an ice sheet only partially constrained by the coastal configuration and which deforms under its own weight
Glaciers constraint by topography	Ice feld	A roughly level area of ice distinguished from an ice cap because of the absence of dome-like form and the control on ice flow exerted by the underlying topography
	Cirque glacier	A small ice mass usually occupying an armchair-shaped bedrock hollow

Contd...

Basic Type	***Component or Sub type***	***Characteristics***
		and characteristically wide in relation to its length
	Valley glacier	A glacier which occupies a rock valley and is overlooked by rock cliffs. They may originate in an ice field or a cirque glacier into which they may imperceptibly merge at their upper end. Large valley glaciers may be joined by tributary glaciers, forming a dendritic pattern of ice flow
	Other small glaciers	Glaciers which occur in a wide variety of topographic positions, but all of which are closely controlled by the underlying topography

'Ice shuts and ice caps are essentially distinguished on the basis of size, the former exceeding around 50,000 sq km in area.

Sources: Based on classification and discussion in D.E Sugdan and S.S. John (1976) Glaciers and Landscape. Edward Arnold, London.

Types of Glacier

Tpye	*Description*
Ice sheet	> 50,000 km^2. Buries underlying relief. Flattened dome in shape.
Ice cap	Dome shaped glacier which buries the landscape. < 50,000 km^2.
ice dome	The central part of an ice sheet or ice cap.

Outlet glacier	An ice stream draining part of an ice sheet or ice cap. It may pass through confining mountains.
Ice shelf	A floating ice sheet of considerable thickness attached to a coast.
Ice field	A relatively flat and extensive mass of ice.
Valley glacier	A body of ice flowing down a rock valley and overlooked by cliffs.
Cirque glacier	A small ice body generally occupying an armchair-shaped hollow in bedrock.
Niche glacier	A small body of ice of an upland lying upon a sloping rock face or shallow hollow which it has modified only to a limited extent.
Diffluent glacier	A valley glacier which diverges from a trunk glacier and crosses a drainage divide.
Confluent glaciers	Two valley glaciers which converge.
Expanded foot glacier	A valley glacier which develops a widened snout where it emerges from confining rock walls.
Piedmont glacier	A large-scale widened glacier formed on a lowland or at a mountain foot. Piedmont glaciers have a flow system which is independent of the valley glaciers and ice falls which feed them.

Glaciers, Non-valley

Ice Sheet: Ice sheet have an area more than 50,000 sq km with a flattened dome which buries the underlying relief. Ice sheets are the largest forms of accumulation and at present there are two such ice sheets—the Greenland ice sheet, which contains 11 per cent of the world's ice and Antarctica, which contains 85 per cent of the world's ice. When ice bodies exceed a diameter of approximately 500 km, the weight of the ice becomes sufficient to cause isostatic down warping of the Earth's crust. Greenland and Antarctica are believed to have subsided by between 25 and 33 per cent of the ice thickness, depressing the rocky surface of central Greenland below sea level. In these, ice sheets radiate out from the centre as a sheet, although near the periphery it may be confined and flow through uplands in outlet glaciers.

Ice Cap: It is a small ice sheet with an area less than 50,000 sq km but which still buries the landscape. Smaller ice caps occur in Arctic, Canada, Iceland and Norway, and may measure only a few tons of kilometres across, for example, Barnes ice cap, Baffin Island ice cap.

Ice Dome: An ice dome is the central part of an ice sheet or ice cap.

Outlet Glacier: An outlet glacier is a stream of ice that drains part of an ice sheet cap and which often passes through confining mountains. They resemble valley glaciers in that they are confined by the topography. Pressure builds up in the ice behind a mountain range and forces outlet glaciers through mountain passes at relatively high speeds—as high as being perceptible. They are common around Vatnajokull ice cap in Iceland and around Greenland and Antarctica. Jacobshavn glacier flowing into Disco Bay on the west coast of Greenland is a good example.

Ice Shelf: An ice shelf is a thick floating ice sheet which is attached to the coast. Thus, there is no friction with the bed and ice can spread freely. The largest example is in Antarctica, the Ross and Ronn- Filschner ice shelves. Ross is larger than France and yet it moves up and down with each tide.

Ice Field: is a relatively flat and extensive mass of ice.

Glaciers' Valley

Valley glaciers are rivers of ice which occupy basins or valleys in upland areas and can be subdivided into—

Valley Glacier Alpine Type: Valley glaciers are long narrow rivers of ice that originate in the snow fields of high mountain ranges and flow down pre-existent stream valleys. They range from a few hundred metres to more than 100 km in length and resemble river stream systems. These glaciers receive an input of water (in the form of snow fall) in the higher reaches of the mountains and have a system of tributaries leading to main trunk system. The flow direction is controlled by the valley, which the glacier occupies and as the ice moves, it erodes and modifies the landscape over which it flows. Valley glaciers are common in young fold mountain areas such as Alps, Himalayas and the Alaskan coastal mountains. The Hubbard glacier, in Alaska is the largest valley glacier being 130 km long. The Siachen, Baltoro, Biafo are such glaciers in the Karakoram mountains.

Cirque Glacier: A cirque glacier is a small ice body that occupies an arm chair shaped hollow in mountains which has been cut into bedrock.

Niche Glacier: A niche glacier is a small upland ice body resting upon a sloping rock face or in shallow hollow that the glacier itself has modified only slightly.

Diffluent, Transaction or Hanging Glacier: These are valley glaciers that have become so thick that they spill over dividing ridges and join with other glaciers in adjoining valleys.

Piedmont Glaciers: Piedmont glaciers are formed where a valley glacier extends over a lowland and spreads out horizontally. For example, the Malaspina glacier of Alaska.

A small glacier such as may develop from a snow patch is called glacieret.

Cold and Warm Glaciers

Based on internal temperature regimes, glacier can be cold or polar ice and warm or temperate ice.

Pure water at sea level atmospheric pressure freezes at 0°C. At greater pressure, relatively lower temperatures are required to produce freezing. Increasing pressure lowers the freezing point of water. At the base of a glacier, 1500 metres thick, the pressure is great enough to lower the melting point by one degree below 0°C. Thus, within this 1500 thick glacier, the ice must become colder and colder with depth to escape melting. So, it must have some internal mechanism to escape cooling. As the glacier grows thicker through accumulation on the surface, a little ice within the glacier melts because of the increased pressure. The heat required for melting can come only from the body of the glacier itself, resulting in the cooling of the remaining ice.

A warm temperate glacier is one which is at the pressure melting point throughout. Warm glacier forms when there is sufficient heat to raise ice temperatures to the pressure melting point. This commonly occurs at the surface where firn experiences surface melting. The surface meltwater sinks into the firn and freezes, if it comes into contact with snow or ice below the pressure melting point. The freezing of the water releases latent heat to raise the temperature of the glacier.

This is such an efficient heating mechanism that on any glacier where firn is subjected to significant summer melting, there is usually enough heat to bring the firn or ice to the melting point during its transformation to ice. In this situation the whole glacier may consist of ice at pressure melting point. Glacier ice at the melting point is in equilibrium with liquid water so water can exist throughout warm glaciers all the way to the base. In such a situation, there is modest decrease of temperature with depth in association with increased pressure.

A cold or polar glacier, in contrast, has a temperature below melting point of ice from top to bottom. Such glaciers get warmer rather than colder with depth and the basal temperatures are lower than pressure melting point. The increase in temperature with depth is caused by geothermal flux. Because a cold glacier is warmer at the base than at the surface, it can conduct the geothermal heat flux to the surface where it gets dispersed into the atmosphere. To do so, the surface of cold glacier, 1500 m thick, has to have a mean annual temperature of about−40°C. So cold glaciers exist only in frigid environments.

A cold based glacier moves largely by internal shearing and deformation and will be unable to abrade the beds. Warm glaciers have a high component of basal sliding in their motion with much greater possibility of erosion and basal deposition. A large glacier need not be totally warm or cold in all its parts. Where thickest, both the Antarctica and Greenland ice sheets are cold towards the surface and warm near the base but in thinner outer parts, they are cold from top to bottom. Also, contrasting thermal regimes may exist in different parts of the same glacier, for example, Antarctica, because of great weight at the centre appears to be warm based, since pressure causes melting but towards its periphery, it is cold based.

As the glaciers move across the surface of a continent, the interaction of a deforming solid with the underlying topography produces a distinctive geomorphological landform.

Glacier Regimes; A World Picture

Region	*Mass Balance*	*Trend*
Antarctica	Accumulation difficult to measure due to drifting but low; estimates 0.1-2.0 cm water equivalent per year. Loss includes iceberg calving, some melting; not so much.	Slight positive mass balance (e.g., 1.1×10^{18} g per year)
Greenland	Accumulation estimates-300mm water equivalent per year; loss normally balances this.	Approximately in balance.
Barnes Ice Cap Baffin Island	Accumulation and loss both small.	Slight overall loss.
North Sweden, e.g., Strongla ciaren	Accumulation (1946-62) 4.0×10^6 m^3 water equivalent; loss 6.2×10^6 m^3 per year.	Loss 55% more than net accumulation; rapid retreat.
Iceland Vatnajokull cap	Considerable variation from year to year but overall loss exceeds accumulation.	Slight loss.
South Norway Nigards-breen glacier	Longer period of observation than most suggests broad periods of advance and retreat in this century	Recent loss (to 1962)
Alaska Malaspina and Upper Seaward glaciers	Variable patterns with occasional surges punctuation retreat phases. Overall excess of loss over accumulation.	Overall loss.

Contd...

Region	*Mass Balance*	*Trend*
Alps, e.g., Hintereisferner	Variable mass balance, but every year negative except one, 1952-61.	Overall loss and retreat 25 m per year.
Southern Alps New Zealand, e.g., Tasman/Franz josef glaciers.	Loss exceeds accumulation west of Southern Alps; variable pattern in this century, but overall excess of loss over accumulation.	General loss.
Equatorial Africa, e.g., Mount Kenya, Kilimanjaro.	Kilimanjaro has accumulation on lower part above snow line and melting near the peak; Lewis glacier (Kenya) retreated 60 m, 1934-58;	General loss.

How Glaciers Move?: That the glaciers move is evident from their velocity characteristics. A glacier's highest velocity is recorded near the centre, the line which diminishes towards the sides where friction against the rock may bring it close to zero. Velocity also tends to decrease with depth, especially in the lower parts of the bed.

Although, sometimes it is difficult to understand how a rigid mass can move and while they do not move like a river, nevertheless, their movement is easily measured.

Glacier Movement

The movements of glacier take place in three ways.

By sliding, over the bedrock; by internal deformation, and by alternate compression and extension.

Sliding Over the Base: Ice normally forms from water at a temperature of 0°C but the temperature at which water freezes is reduced under pressure, and as a glacier moves, it will exert pressure, and therefore, some melting may take place at its base. Thus, a thin film of water exists between the glacier and the bedrock. This film reduces friction, and hence, allows the glacier to slide. This film is likely to occur in warm glaciers. Expressed as percentage of the total movement basel slip varies from 10 to 75 per cent with a mean value of around 50 per cent. The presence of a water film at the base of the glacier is an important requisite and at least a portion of the movement probably takes place by the refreezing of water (regelation) around small irregularities in the bedrock floor. This occurs because there is increased pressure on the upstream side of any protruberance, which leads to local melting around that area. Then the water flows downstream where, it refreezes under conditions of increased pressure.

Internal Deformation: Ice is quite brittle and at the same time prone to deformation. If a horizontal bar of ice is left unsupported at one end, it will deform under its own weight. In case of glacier ice, deformation or creep will take place under the action of gravity depending on the stresses involved—the weight of the overlying ice column and the slope of the upper ice surface. Thus, the rate of plastic deformation will increase with increasing surface slope. Internal temperature regime also plays an important part, for example, warm ice moves more easily than cold ice.

Thus, glaciers in the temperate zone will move faster on steep slope than a thin flat ice cap in polar areas.

Alternate Compression and Extension: Under some conditions, compression and extension take place because ice cannot deform sufficiently quickly to the stress within the ice. As a result, fractures develop in ice and movement takes place along a plane.

Velocities and Directions of Flow: Measurements of surface velocity across a valley glacier show that surface ice in the central part of the glacier moves faster than ice at the sides, similar to the velocity distribution in a river. The reduced rates of flow towards the margins are due to frictional drag against the valley walls. A similar reduction flow rate towards the bed is observed in a vertical profile of velocity.

Although, snow continues to pile up in the accumulation area each year while melting removes snow and ice from the ablation area, the surface profile of a glacier does not change much because ice is transferred from the accumulation area to the ablation area. The dominant flow is downward in the accumulation area where mass is being added, and upward in the ablation area where mass is being lost.

Crystals of ice that enter the glacier near its head, therefore, have long path to follow before they emerge near the terminus. Those falling closer to the equilibrium line, however, travel only a short distance through the glacier before reaching the surface again.

Flow velocities in most glaciers range from only a few centimetres to a few metres a day, or about the same slow rate that groundwater percolates through crustal rocks. Hundreds of years have elapsed since ice, now exposed at the terminus of a very long glacier, fell as snow near the top of its accumulation area.

Glacial Tide

Some glaciers move at velocities, much higher than normal, for short periods and, are thus, said to surge. Whereas normal glaciers flow at 3-300 m $year^{-1}$ surging glaciers may flow at rates of over 10 km $year^{-1}$. Glaciers in Himalayas and the United States are known to have moved at rates over 300 m per day at the height (or peak) of a surge. In such cases, the jump from normal to surging flow takes place very suddenly. Hassanabad glacier, Karakoram surged 10 km in 2½ months, Kutian glacier also in Karakoram surged 12 km in three months, Sustina glacier in Alaska is also prone to surging.

Some glaciers flow permanently at surging velocities, for example, Jacobshavn Isbrae in west Greenland nourished by Greenland ice sheet flows at 7-12 km $year^{-1}$.

Glacial surges are supposed to be caused by an instability that is induced when snowfall that collects in the accumulation zone is not transmitted efficiently..Instead, there is a prolonged storage of surplus snow and ice, which causes the glacier to grow in bulk to unstable proportions. Finally, the glacier surface becomes so steep that it is out of equilibrium. Once some critical threshold is reached and some triggering mechanism (earthquake) has been there, the glacier begins to surge.

Glacial surges are very dangerous. The advancing ice front may cause damage to farms. Large quantities of meltwater may be released causing floods. If surging glacier enters the sea, dangerous icebergs create hazards for shipping, it may block a stream creating natural lake behind.

Glaciers produce many features. These features are divided into two main groups:

1. Those produced by erosion.
2. Those produced by deposition.

Glacier's Erosion

Erosional features are produced as a result of the movement of glaciers.

Mechanics of Glacial Erosion: There are two ways in which glacier erodes:

Glacial Plucking: It is the lifting and removal of fragments of bedrock by a glacier. It is one of the most effective ways in which glaciers erode the land. The bedrock is loosened by the freezing and thawing of water in fractures beneath the ice.

When the glaciers melt or when there is rainfall, this water seeps down along the sides of the ice mass filling up the cracks, fissures and pore spaces within the rock along the edges and also at the head of the glacier. When the temperature drops below freezing point, water in the cracks, pore spaces and fissures expand.

The ensuing increase in volume exerts enormous pressure on the adjoining rocks and breaks them apart. The loosened blocks freeze to the bottom of the glacier and are plucked or quarried from the bedrock and then get incorporated into the moving ice. This process is especially effective where the bedrock is cut by numerous joints and where the surface is unsupported on the downstream side.

Abrasion: The angular blocks, plucked and quarried by the moving ice, freeze firmly into the glacier where they act as tools that grind and scrape the bedrock. The overlying ice pressure makes the angular fragments very effective agents of erosion, capable of wearing away large quantities of bed rock. The rock fragments incorporated into the glacial ice are themselves abraded and worn down into peculiar faceted stones by grinding against the bedrock surface.

Besides these two modes of erosion, mass wasting plays a very important role in erosion of glacial regions. Along the margins of valley glacier, the valley sides are scraped and blocks are broken off which become frozen into ice and are carried away. This leads to undercutting of the sides of the valley.

Once a side is undercut, the probability of mass wasting is greatly increased. Slumping, sliding and debris avalanches bring huge quantities of debris on top surface of the glacier.

Factors Affecting the Mechanisms of Glacial Abrasion

Fundamental Factors	*Comments*
Presence of debris in basal ice	Clean ice is unable to abrade solid rock. The rate of abrasion will increase with debris concentration up to the point where effective basal sliding is retarded.
Sliding of Basal Ice	Ice frozen to bedrock cannot erode unless it already contains rock debris. The faster the rate of basal sliding, the more debris passes a given point per unit time and the faster the rate of abrasion.
Movement of debris towards glacier base	Unless particles at the base of a glacier are constantly renewed, they become polished and less effective abrasive agents. Thinning of the basal ice by melting or divergent flow around obstacles brings fresh particles down to the rock-ice interface and increases abrasion.
Other factors affecting nature and rate of abrasion	
Ice thickness	The greater the thickness of overlying ice, the greater the vertical pressure exerted on particles on the glacier bed and the more effective is abrasion. This is the case up to a depth where friction between particles and the bed becomes so high that movement is significantly retarded and abrasion decrease.
Basal water pressure	The presence of water at the glacier base, especially when at high pressure, can reduce the effective pressure on particles on the bed, and thus, abrasion rates by buoying up the glacier. However, sliding velocities may tend to increase because of the reduced friction.
Relative hardness of debris particles and bedrock	The most effective abrasion occurs when hard rock particles in the glacier base pass over a soft bedrock. If the debris particles are soft in comparison with the bedrock, the former are abraded and little bedrock erosion is accomplished.
Debris particle size and shape	Since particles embedded in ice exert a downward pressure proportional to their weight, large blocks should abrade more effectively than small particles. Moreover, angular debris will be a more efficient agent of abrasion than rounded particles.

Efficient removal of fine debris	To sustain high rates of abrasion, fine particles need to be removed from the ice-rock interface since they abrade less effectively than larger particles (assuming the latter are continually being supplied from above). Meltwater appears to be the main mechanism for the removal of fine (10.2 mm) debris.

Source: Based on discussion in D.E. Sugden and B.S. John (1976) Glaciers and Landscape. Edward Arnold, London, pp. 153-5.

Main Characteristics of Erosion

	Valley Glaciers	*Continental Glaciers*
	Small Features	*Small Features*
1.	Striations	Striations
2.	Grooves	Grooves
3.	Polished surfaces	Polished surfaces
4.	Crescentic marks	Crescentic marks
5.	Crescentic gouges	Crescentic gouges
6.	Chatter marks	Chatter marks
7.	Glacier Mills or Moulins	Glacier Mills or Moulin
8.	Glacier Tables	Glacier Tables
	Large Features	
1.	Nivation hollows	A general suppression of whatever topography existed before glaciation
2.	Cirques	
3.	Tarn Lake	
4.	Arete	
5.	Horn	
6.	Col	
7.	U-shaped valley	If continental glaciers covered, many of the features of valley glaciers may be formed.
8.	Glacial trough	
9.	Glaciers stairway	
10.	Paternoster lake	
11.	Truncated spur	
12.	Hanging valley	
13.	Ribbon Lake	
14.	Roche Moutonees	
15.	Crag and Tail	Nunatakas
16.	Fiords	
17.	Fiards	
18.	Tunnel	

Striations: Striations are simply scratches on rock surfaces. If the glacier carries gravel or boulders, the rock will be scratched and if the rock is softer, than the fragments, bed rock will be striated.

Polished Surfaces: If the under surface of the glacier contains sand and silt, these act like sand paper, smoothening and polishing the rock surface.

The grinding action of the moving ice further crushes the grains to produce an abundance of fine particles known as rock flour.

Grooves: Large scale abrading fragments, such as gravels and boulders caught within a moving ice, develop striations of a larger size, a metre or more. These are known as grooves or flutings. Grooves or flutings are excavated across varying lithologies and geological structures. Flutings can individually be up to 20 km long, 25 m high and 100 m wide.

Friction Cracks: Friction cracks are arcuate in form, either concave or convex, in the direction of movement created as a result of transmission of stress, generated by the moving ice. The combination of high overburden pressure and rapidly moving ice generates enough stress to fracture brittle bedrock. The most common glacial friction cracks are crescentic fractures, lunate fractures and crescentic gouge.

Sickle troughs are produced where ice blocks, while moving, hit obstructions and chip-off pieces of bed rock.

Glacier Mills or Moulins: When the sun is shinning and the temperature is somewhat higher, small pools and rills develop, finally gathering into streams which fall into crevasses. Thus, melting and potholing wear deep couldrons through the fissured ice called glacier mills or moulins. The water may escape to the snout through a tunnel.

Nivation Hollows and Cirques: Glaciers do not form unless there are suitable hollows in which snow can pile up to sufficient depths. Such hollows are commonly prepared by snow patch erosion or nivation and are called nivation hollows.

Snow banks commonly linger longer in the summer on the flanks of lofty mountain peaks. Some snow banks may be perennial, occupying the same spot year after year. During summer, the melting snow bank keeps the area beneath it and immediately around it well supplied with water, especially down slope. With the help of water, weathering and erosion proceed very effectively in the vicinity of snow banks than in snow free areas. In autumn and spring and sometimes in summer, night temperature falls low enough to refreeze some of the day time water. Repeated freezing and thawing break rocks into finer particles and cause them to creep down slope.

By this, means of nivation, i.e., the localised erosion of a hill side, by frost action and mass wasting at the edges and beneath lingering snow patches, a hollow is created. This is a nivation hollow. The broken rock fragments from the top of the backwall of the nivation hollow fall on the top of the snow bank and slide downwards to its lower edge, forming a linear or curvilinear ridge having a composition similar to a moraine. This structure is known as a protalus rampart.

Nivation hollows are prerequisite for ultimate glacier development. As the hollows grow so does the snow bank, gradually increasing in thickness, as well as, in area. Eventually, it becomes thick enough so that the lowest layers are converted to ice. When the mass attains thickness of 60 to 100 metres, it starts to move by internal deformation and by sliding over its bed. At that time, it becomes a small glacier. Owing to movement, erosion is greatly increased and soon the small glacier, excavates basins on the side of the peak.

The debris derived from weathering get accumulated along the down slope edge of the ice, building an embankment which further enhances the basin like form of the depression. The little ice mass in this basin now becomes a cirque glacier. The glacier would start plucking and quarrying at its head as a result of downward percolation of water and the frost shattering of the rocks in the headwall, supported by strong solar radiation and large diurnal and seasonal changes in high altitudes. Draining into cracks and joints, the water freezes and breaks up the rocks until the ice carries them away. The cirque glacier, thus, literally eats backward into the bounding slopes, undercutting them so that they become oversteepened and contribute to the enlarging process by shedding coarse debris into the glacier. This headward process is called sapping. As this headward sapping proceeds, a steep wall is created. The rotational slip of the glacier deepens the cirque floor, leaving a rock bar at the mouth. These steep sided semi-circular bowl shaped depressions are called cirques. They have different names in various regions. For example, they are known as cirques in Alps, corries in Scotland, cwms in Welsh, kar in Germany, nisch in Sweden and bothkjedel in Scandinavia.

Some cirques have been so overdeepened by glacial excavation as to have a topographic closure as deep as 100 metres behind the bedrock sill. These basins contain sparkling, crystal clear, cirque lakes. Tarn lake is the local name given to a small lake in N. England. These are often but not necessarily a cirque lake.

In middle latitudes, cirques occur in groups at relatively high altitude and above the level of major glacier valleys, at the head of glacier system or along the top of valley sides. In and around Antarctica, they may occur just above sea level. Cirques are most numerous where the amount and rate of accumulation could give the greatest intensity of erosion, i.e., where the alignments would permit surface shade from melting sunshine and where snow accumulation and drifting would be encouraged. Cirques in Northern Hemisphere are on the north facing slopes and in the Southern Hemisphere on south facing slopes. The areas, which have longest periods of glaciation, have the best developed cirques, for example, in Antarctica, High Himalayas and Alaska.

Arete and Horn: As cirques develop, they eat back into the hill mass by headward sapping. When several cirques lie close to one another, the divide separating them may become progressively narrowed until it is reduced to a narrow precipitous knife-edged ridge bounded on either side by the steep head walls of the opposed cirques. This is known as an Arete in France and Grat in Germany. If the bedrock is homogenous, the skyline profile of an arete will be smooth, but if the rock is heterogenous and irregular in structure, particularly if it is jointed, the arete profile will be rough and jagged, in other words, like a saw-toothed ridge. Spires and joint-defined columns of rock, standing on its crest, are called gendarmes. If the glaciers whittle away the mountain from all sides, the result is the formation of a pyramidal peak known as horn, of which the Matter horn of Europe is the best known example.

Col: When two cirques converge cutting into the same wall, the divide wall is lowered by hundreds of metres forming a saddle-shaped gap called Col. Cols are also produced by the glacial breaching by diffluent glaciers. When the ice cannot get away from a valley fast enough (perhaps its valley is blocked lower down by other ice or because there is constriction), it will over flow at lowest available point. The result of this erosion is the creation of a col or gap in the watershed. This is exhibited by the Rimu glacier of the Karakoram.

U-shaped Valley: U-shaped are valleys so called because of the shape and are formed since erosion occurs throughout the base of the glacier, which becomes progressively less at the sides and is not being concentrated at a point as it happens in rivers.

Glacial Troughs and Associated Features: Well developed valley glaciers are called Glacial troughs. Glacial troughs have a steep headwall known as trough end, which closes off the upper end of an over deepened glacial valley around which a number of cirques frequently occur. The head of the glacial trough is not a cirque headwall but the lower edge of cirque threshold. There is usually a conspicuous drop from a cirque threshold, to the floor of a glacial trough, called trough headwall. The glacier ice from the surrounding cirques (which act as tributary), coalesce and feed into the main trough. This greatly increases the power of valley glacier by increasing its volume and morainic material thereby leading to more effective down wearing of the valley head.

Valley glaciers tend to occupy valleys which were originally created by river systems before glaciation but valley glaciers have so altered them that neither in cross profile nor in long profile do they resemble stream carved valleys.

Ice is not like water which can get easily diverted by obstructions. The digging and truncating action of the glacier on the sides of the pre-glacial action will be proportional to the mass and weight of ice. After some time, the V-shaped pre-glacial valley is changed to a U-shaped form. The ground (base) and the walls are plucked, and subsequently, removed creating a new steeper slope. Thus, in successive stages the V-shaped valley is transformed into glacial trough.

Glacier troughs occur in a variety of situations:

1. The Alpine type are contained in pre-glacial stream network, separated from other troughs by high ground.
2. The Icelandic type of trough is associated with erosion beneath ice caps and ice sheets. It is nourished in its headward region by steep ice falls from adjoining plateaus.
3. Composite troughs include sections of pre-glacial valleys and also sections of trough formed by transection of glacier breaking through former drainage divides.
4. Intrusive troughs occur where ice was thrust against a highland area after a lowland path. This occurred in finger lake area of New York State.

The longitudinal profiles of glacial trough are usually quite irregular, characterised by rock basins, rock steps, frequent breaks in slope, and paternoster lakes, chains of small lakes connected by inlet and outlet streams.

Glacial Basin

Glacial troughs possess rock steps, rock basins, roche moutonees, glacial stairway, truncated spurs and hanging valleys.

Rock Steps: There is very uneven excavation of the glacier floor depending on the nature and structure of bedrocks, thickness of the glacier and the rate of glacier flow. Poorly consolidated strata are scoured out more rapidly than resistant rocks and well jointed tracts are quarried away by plucking. Thus, when the glaciers come across a rock sequence, which is varying in resistance, the floor is excavated into a series of successive steps called rock steps.

Rock Basins: Rock basins are sections of the valley, overdeepened by the glacier by either rotational movement or by selective erosion of highly jointed and fractured bedrock by the passage of glacier ice. The size of rock basins varies and are oriented in the direction of the flow. Such rock basins are now occupied by lakes called trough lakes, finger lakes or ribbon lakes. The riser at the end of each basin or moraine, forms a dam. In the long profile, the glacial troughs may exhibit a series of basins separated by resistant edges of rocks. These basins form chains of lakes, each with a waterfall over the downstream ledge of the next lake. These chains of lakes are called paternoster lakes (likeness to paternoster or string of rosary beads).

Glacial Stairway: The step like longitudinal profile of some glaciated valleys resembles a giant staircase, called glacial stairway or cyclopean stairs in which relatively flat treads or basins are separated by steep risers and associated reigel.

A riser marks the down valley end of each step and the ascent to it is from the step below. A riegel is a sort of rock bar at the top of and just back of a riser. When there is a series of bars, glacial stairways develop. A tread is relatively flat floor of a step. A tread may even be hollowed into a basin.

Cyclopean stairs may form in any of several different ways:

1. Differences in resistance to glacial erosion, especially with respect to frequency and spacing of joints. The sub-glacial floor may be differentially eroded because of these differences. For example, more rapid erosion in highly jointed granite susceptible to plucking can selectively etch out basins, whereas more massive granite is more resistant and forms steep risers.
2. Discordance of sub-glacial floors at tributary junctions. Increased sub-glacial erosion due to glacial thickening below the confluence of tributary valleys deepens the valley floor successively below each junction.
3. Back wasting of cirque head walls at successively higher elevations at the heads of glaciated valleys. Progressively rising snowlines result in abandonment of lower cirques while head wall recession continues to steepen slopes at the heads of valleys.

Since there are differences in processes related to these mechanisms of formation, not all cyclopean stairs can be attributed to a single type of development. Each occurrence must be evaluated on the basis of its own evidence.

Truncated Spurs: Truncated spurs are the steepened bluff on the side of a glacial trough, in between tributary valleys. Most of the glacial valleys were river valleys before. Since glaciers are not as easily deflected as rivers, after meeting any obstacle, the glacier flow is more or less streamlined. Thus, if the glaciers were flowing in interlocking spurs, the sharp and acute bends occurring along the course of streams are more or less straightened up by glacial abrasion and it cut the projecting spurs to produce blunt triangular facets known as truncated spurs.

Hanging Valleys: When rivers create a drainage system consisting of a trunk with a tributary stream, each tributary does it best to make on accordant junction with the trunk at a common level without a break in the longitudinal profile of either. In a discordant junction, one of the valley floors, nearly always the tributary, lies or hangs distinctly over the other. The difference can be hundreds of metres. One such manifestation of such a discordant junction is the Hanging valley.

Where the tributaries enter the main glacier, the upper surfaces of the glaciers are at the same level. The main glacier with a larger volume of ice and a greater thickness is able to erode its valley to greater depth than the tributary glacier. Thus, tributary valleys have their lower ends cut clean away as the spurs between them are ground back and truncated. When the valley glacier has finally receded, the flows of the tributary valleys will be higher than the main valley, and thus, the tributary valleys are called Hanging valleys. The difference in level can be hundreds of metres. These hanging valleys often become site of waterfalls, some of which are among the highest in the world. The Yosemite falls in California (Sierra Nevada) is a renowned example.

Fiords vs Fjords

Fjords or Fiords are long narrow inlets into the sea coast with more or less steep sides. Fjords form when glaciers making their way to the sea, scooped out deep trough like valleys (glacial troughs). When these troughs were submerged due to eustatic changes, the lower end of these troughs were filled with water and became inland stretches of sea. Thus, they differ from land based glacial valleys in that they are submerged under sea. Some of these fiords are extraordinarily deep, the maximum known depth of Greenland, Norwegian and Chilean fjords are of the order of

1300-1400 m and in Antarctica fjords, almost 1000 m deep, are known. These great depths are usually separated from the ocean by a sill of solid rock, over which the depth is only around 200 m or so. It is also thought that these are submerged moraines. They may also be the result of greater glacial erosion up fjords where the ice was thick and actively eroding and less erosion in terminal zones where the ice was too thin for effective erosion. Fiards differ from fiords in that they are shorter and shallower. The difference between the two is illustrated in Norway and Sweden. Norway is a land of fjords, Sweden has less fjords but many fiards. The same feature is known as forden in Germany and firths in Scotland.

Fjords and Fiords should not be confused with fohrde which is a German term for the shallow estuaries of the drowned linear valleys in the lowlands around southern Baltic.

Streamlined Features: Recently deglaciated regions, commonly exhibit many types of elongated, streamlined forms of all sizes and gradations from entirely bedrock to entirely unconsolidated deposits. The streamlining of material beneath a glacier minimises the resistance of knobs and hills to glacial flow. Such types of streamlined features are mostly found in southern Finland where thousands of smooth, elongated hills give the landscape a distinct linear aspect.

Individual streamlined rock knobs are generally a few metres high and 10 to 20 m long, usually with one end higher than the other. The three well-known streamlined features are whale back, rock drumlines and roches moutonees. Whale backs are bedrock features that occur singly but are more common in schools.

The name relates to their smoothed, somewhat elongate, gently curved form resembling the back of a whale rising above water level. Where small, they might be more properly called dolphin backs or porpoise backs. Whale backs are mostly 5 to 10 metres long, 3 or 4 metres wide, and a metre or two high. Sizes can be mixed within a school. They usually display modest fore and aft symmetry with the stops side a little blunter and steeper. The surface is usually striated and often grooved and polished. Whale backs taper in the down glacial direction and grade into rock drumlins.

Rock Drumlins: **Rock drumlins are hillocks, moulded or eroded into a streamlined form by glaciers.** Like whale backs, rock drumlins are smoothly abraded, but they are larger than whale backs with lengths measured in many tens to 100 metres, and their asymmetry is more marked. Rock drumlins have a steeper blunt-nose facing upstream and a narrower, gentler tail extending downstream, forming an elongate teardrop shape as viewed from the top. These form as some bedrock hillocks have been eroded into a streamlined drumlinoidal forms by moving ice.

Roche Moutonnee: Roche moutonnees are a good example of bedrock landscape feature formed by abrasion and plucking combined. A roche moutonnee is prominent bedrock knob, rounded, smoothed, grooved, striated and locally polished by abrasion on one flank, and made steep, jagged, cliffy and irregular on the opossing flank by plucking.

Each flank is impressive in its own way. The smoothed, gentler flank provides the name, a moutonnee. The configuration is supposed to resemble the smoothed flank. It faces upstream and, is thus, the stoss side. Some plucking has probably occurred on this flank, but plucking there is overshadowed by abrasion. By contrast, scant signs of abrasion are found on the lee side of the knob; plucking has completely dominated. Blocks of plucked rock are strung out downstream from some roche moutonnee. In addition to being steeper, the plucked face is usually also higher because the area leeward of the knob has been lowered through excavation by plucking. The lateral sides of the knob are usually smoothed, abraded and grooved, so abrasion features may be seen over about two-thirds of its surface.

Roche moutonnees range from knobs with dimensions measurable in a few metres to major prominences 50 to 200 metres high and two or three times as big in width and length. Many are elongated in the direction of ice flow.

Roche moutonnees inhabit areas in which glaciers were in a vigorously eroding mode. They favour resistant but jointed rock. Most of them probably started as a pre-existent knob, created by pre-glacial erosion that was modified by glacial action. Small roche moutonnees may have been created solely by glacial erosion. The larger form of roche moutonnees is known as flyggberg.

Crag and Tail: Crag and tail are the result of the presence of an obstructive mass of rock that lies in the path of oncoming ic This mass protects softer rocks in its lee from the effects of glacial erosion for the ice appears to have moved over and around the crag, leaving a gently sloping 'tail' in its lee.

Tunnel Valleys: Glaciers produce a vast amount of meltwater, some of which flows on glacier, some within and some over the bedrock, i.e., subglacially (under) and englacially (within). This water flows at a rapid speed before emerging to flow along the side of the ice or down valley away from the glacier snout. This water, laden with sediments and flowing under great hydrostatic pressure, carves out steep sided flat floored valleys called tunnel valleys or meltwater channels.

Nunataks: Where, as the erosive and depositional forms of valley glaciers are largely confined to the valleys through which they move, an ice sheet mantles and modifies almost all land surfaces. Only occasionally and usually on the margins do high peaks project through the ice shelf as nunataks.

Terminology of Periglacial

The term periglacial (next to ice) was introduced in 1909 by Lozinski to refer to the area bordering the continental Pleistocene ice sheets and is characterised by severity of its climate. Since then, the term has been extended in a number of ways (sometimes contradictory) to unglaciated Polar and Alpine environments, in general, or to the impact of geomorphologically effective soil and rock frost, in particular. Periglacial, now is generally applied to a particular form of climate, involving virtually tundra vegetation and permanently frozen subsoil, and the geological and geomorphological conditions that go with those conditions.

The earlier definition considered it simply in terms of proximity to glaciers and ice sheets and did not involve identity of climate. It is quite possible for vigorous confluent glaciers from high mountains to reach down into the coniferous forest zone of the surrounding areas where the climate is certainly not periglacial. At the same time, there are large parts of Siberia which are far from glaciers and ice sheets (as they are arid lowlands) but which are truly periglacial in climate. Although the periglacial concept is somewhat imprecise and broad enough to defy quantification, Chorley, Schumm and Sugden have defined a periglacial morphogenetic region. It is characterised by a mean annual temperature of 0°C to -10°C and an average annual precipitation from 0 to 1500 mm. Ideally the periglacial region consists of the zone of perennially frozen subsoil (permafrost), seasonally thawed topsoil (active layer), frequent changes of temperature across the freezing point, and an incomplete vegetation cover of herbaceous plants and dwarf species of trees. However, for practical reasons, the periglacial zone is regarded as coincident with the region of permafrost and active layer.

Periglacial Areas: Atmosphere

Periglacial areas exist where some broad environmental conditions are present. In case of high latitude periglacial realm, these conditions are:

1. Long and cold winters with protracted or daily frost for ten months of the year. The snow cover is normally shallow and frequently incomplete. This slows the seasonal frost to penetrate deeply into the soil.
2. The summers are modest and the subsoil temperatures at a depth of 10 cm hardly rise to more than 5°C-10°C even when the ground surface temperature are 24°C or more. This prevents any chemical weathering except carbonation which is active as long as temperature remains above 0°C.
3. As a result of short summers, the annual thaw is commonly inadequate to melt the winter frost. A large portion of the subsoil remains permanently frozen (permafrost) above which is the thaw zone. The thaw zone is variable in depth ranging from an (5 cm) in boggy soils to as much as 4.5 m in well drained sandy soils. Beneath this thaw zone, the permafrost layer may have a thickness of tens, hundreds or thousands of metres.
4. Permafrost development impedes soil drainage in the summer when there is melting. The release of enormous amount of water leads to year round water logging with widespread development of lakes and swamps.

5. Deep frost facilitates active mechanical weathering in the subsoil as well as frost shattering of the rock outcrops. As the soil thaws down in the autumn, there is produced mixing, churning or heaving of the regolith. Mixing takes place because of the differential conductivity and expansion of stones compared with those of wet, fine, soil aggregates. The formation of ice crystals underneath stones and their consequent growth into ice needles pushes stones towards the surface replacing the thaw ice by fine soil. Similarly as frost penetrates deeper, pressure from beneath causes the tearing up of frozen subsoil.

Characteristics of Glacial Movement

Glaciers, themselves are characterised by certain features, the most important of which are crevasses, seracs and pressure ridges. These are neither erosional nor depositional features, rather they result from some internal mechanism.

Crevasses: Crevasses are cracks of variable width in the surface ice of a glacier, caused by shear stresses set up by differential movement within the ice. Crevasses are of various types – Marginal Crevasses result from differential drag against the valley sides. These are formed by stretching, produced by large differences in flow velocity, along the glacier's margin owing to the drag of the valley walls. Longitudinal Crevasses are roughly parallel to the direction of the flow from where ice spreads laterally. Transverse crevasses develop across a glacier wherever there is marked convexity in the shape of the floor. In this case, the whole thickness of the glacier is broken into blocks which may reunite and flow. Radial crevasses are formed when the glacier spreads out. Bergschrund is a special type of crevasse which is very wide and deep. This opens in summer near the top of the nevee field of a cirque where the head of a glacier is pulled away from the precipitous walls from ice adhering to them. Several such features may open instead of a large one.

Crevasses, however, are only temporary features. Some are closing while others are opening.

Pressure Ridges: Pressure ridges are formed when a glacier is subjected to lateral confinement while passing through a constricted course.

Seracs: Seracs are the upper part of the broken glacier, which are jumbled mass of jagged ice pinnacles.

Lines that are Trim

When an ice stream moves down a forested valley and then recedes, it leaves a knife sharp line between the barren deforested ground and standing timber. This is a trimline. With time, vegetation springs up in the deforested area but the difference in tree size is quite obvious and the line remains well defined. A line marked by trees of different ages might be termed as a palaeotrim line. The oldest trees growing inside the trimline provide a minimal date for the glacier advance. The exact date can be determined by locating a living tree along the trimline that was tilted but not runover and killed by glacier advance. The rings of tilted tree trunks grow eccentrically. By counting the eccentric rings back to the time of tilting, the date of glacial advance can be known.

The Channels

Once entrained, glacial debris is transported in three main ways: as surface (supraglacial), internal (englacial) or basal (subglacial) material. Debris falling on to the glacier from the valley walls commonly forms a long ridge of material at the glacier edge, known as lateral moraine. Where two or more ice streams merge, these moraines may coalesce in the centre of the glacier to form a medial moraine. Similar features develop where ice diverges and then rejoins around a rock outcrop.

Debris also reaches the surface through upward shearing of the ice and by down-wasting which releases englacial material. These sediments typically accumulate as a thin layer of sediment which acts to insulate the underlying ice. As a result, beneath the debris, summer melting is inhabited and a small mound or cone of ice is produced, veneered with sediment. This is known as a dirt cone. Eventually, the sediment washes or slips off the cone and the exposed ice melts.

Englacial debris originates from a number of sources. Some reaches the ice interior along shear planes. Other material is washed into the ice down crevasses. Over time, some material works its way downward through the ice as a result of localised melting and regelation while debris may also be buried by overriding ice. Large volumes of englacial debris are concentrated by meitwaters in ice tunnels.

The basal debris of a glacier comprises both material frozen into the ice and debris which is being dragged along beneath the ice as a result of tractive forces. Traction of subglacial material is particularly effective where the ice is crossing relatively loose sediments or plastic bedrocks such as clay and may extend to a considerable depth. The effects are shown by deformation of the underlying strata into complex folds.

Glacier's Deposition

Glacial deposition involves a range of processes and any classification is somewhat arbitrary. An added complication is that researchers have used different terms to describe the same process. Here, we make a broad grouping of processes on the basis of whether they operate at the base, on the surface, or around the margins of a glacier. Nevertheless, material deposited in one situation may be subsequently re-entrained and redeposited elsewhere.

Classification of Mechanisms of Glacial Deposition

Location Relative to Glacier	*Process*	*Thermal and Dynamic Conditions*
Subglacial	Undermelt	Warm-based only, active or stagnant
	Basal lodgement	Predominantly warm based, active only
	Basal flowage	Warm based only, active or stagnant
Supraglacial	Meltout	Active or stagnant
	Flowage	Active or stagnant
Marginal	Dumping	Active only
	Pushing	Warm or cold based, active only

At least three distinct mechanisms of subglacial deposition can be identified. Undermelt involves the deposition of material through melting of the underlying ice. Melting may arise from geothermal heat, through frictional heating or as a result of increased pressures around the up-glacier side of obstructions. A second process, called basal lodgement, entails the smearing of predominantly fine material on the glacier bed. Lodgement of subglacial debris occurs where the friction between a particle being transported by the ice and the glacier bed becomes so great that its further movement is retarded. This situation is favoured by thick ice promoting high basal pressures which increases friction on the glacier Bed and by low flow velocities which are associated with weak basal shear stresses. Basal flowage is a third mechanism of subglacial deposition but it is also partly erosional. It involves both the squeezing of unconsolidated water soaked debris into basal ice concavities and the streamlining of till by overriding.

Supraglacial debris deposition can occur by melt out or by flowage. Melt out involves deposition of sediment through melting of the glacier surface. It is most active in the snout of warm glaciers where up to 20 m of ablation may occur in a single summer. Supraglacial flowage, involving the movement down the ice surface of the debris-rich upper layers of glacier, has also been observed and is particularly common near the snout where intense supraglacial melt out occurs. The movement ranges from slow creep to a rapid liquid flow.

Deposition around glacier margin can arise from a variety of mechanisms. Water-soaked till can be squeezed from under the ice, and if the glacier margin remains stable for some time significant accumulations of supraglacial and englacial debris can form through dumping of material by melt out. Existing glacial deposits can also be pushed by an actively advancing glacier and deposited when the advance stops, or be overridden and recycled towards the snout as englacial and supraglacial debris. Material deposited at a glacier margin commonly suffers extreme disruption, prior to deposition, so that any original orientation of particles in the ice is usually lost, except possibly where a frozen mass of till is bulldozed *en masse*.

Because of the wide diversity of glacial depositional environments, the variety of landforms produced by glacial deposition is almost endless, with many transitional forms. Depositional landforms may be grouped into categories related to the following: —

1. Position of deposition from the glacier (subglacial, ice marginal or proglacial)
2. Relative vigour of ice motion (stagnant ice or active ice)
3. Morphology and composition.

Although, most depositional glacial landforms are characterised by one or more of the sediment types discussed above, the landforms are named largely, but not completely, on the basis of their shape and size rather than on the material composing them.

Till Deposits: The unstratified, unsorted debris dropped more or less randomly by the glaciers is called till.

Deposits of Active Ice: In actively flowing glaciers, sediment transported by the ice is plastered on to the ground as till or is released by melting at the glacier margin where it either accumulates as a moraine or is reworked by meltwater and transported beyond the terminus.

Till: Traditionally, two types of till have been recognised:

Lodgement Till: It is laid down subglacially when debris is released directly from the sole of the ice. At its simplest, this involves pressure melting of basal so that debris in it is released and then lodged or plastered on to the glacier bed. Lodgement may occur beneath stagnant ice, but it certainly occur beneath actively moving ice and hence, some of the till has the water squeezed out of it, and it may be sheared and made brittle by the pressure of the overriding ice. During this process, alignment of the long axes of particles occurs. Lodgement tills can be built up to a thickness of many metres by progressive melting out from the glacier bed and by shearing of one layer of till over another.

Lodgement till may be shaped into a variety of landforms. Where the ice is relatively thick and actively flowing, extensive, more or less flat till-plains may form. Differential stresses within the ice, however, mould the till into streamline features such as flutes and drumlins. Flutes vary in height from a few centimetres to several metres and may be between a few metres and many kilometres in length. Exactly how they form is not clear, but it appears that they are often initiated by boulders which protrude into the ice base and produce an elongated ice cavity. Material is squeezed laterally into this cavity by the weight of the ice to form a small ridge.

Ablation Till: It accumulates initially in a supraglacial position and is later lowered to the ground surface by melting. Ablation till consists of:

1. Melt out till is the direct product of ablation continuing beneath a cover of detritus.
2. Flow till consists of debris that has built up on the ice and after saturation with meltwater. It becomes so unstable that it flows or slumps into nearby hollows. Till is deposited as series of distinctive landforms.

Characteristics of Till

1. *Poor sorting:* There is a large ingrain size, with large clasts up to boulder size frequently contained in a finer, sometimes clayey, matrix.
2. *Lack of stratification:*. Laminations and graded beeding (progressive change in grain size with depth) are generally absent except in deposits modified by meltwater which may exhibit stratification.
3. *Mixture of lithologies:* Particles may have been derived from widely separated sources, especially in the case of deposits laid down by large ice sheets.
4. Frequent presence of particles with abraded facets and striations.
5. Preferred orientation of particles.
6. Compaction associated with pressures developed during deposition.
7. Overlies striated rock or sediment basement.
8. Predominantly sub-angular particles due to combination of fracturing and rounding by abrasion.

The Grounds

Moraines consist of materials which are heterogenous in shape and size and include angular blocks of rock boulders, pebbles and clay that have been transported and deposited by glacier or ice sheet. Classification of moraines are based on the mode of their formation, i.e., in a supraglacial, subglacial or englacial environment. The great variety of linear moraines, some of which exhibit sorting by meltwater on their outer slopes where glacio-fluvial processes have been active, include:

Lateral Moraines: These are ridges of glacier debris on the margin of a glacier or along the sides of a valley, formerly occupied by glacier. When ice is present, lateral moraines may bury the glacier edge in which case the debris protects the ice from surface melting.

Medial Moraines: It is a linear accumulation of heterogeneous material extending down the centre of a glacier. It is formed by the merging of two lateral moraines from the point where two glaciers unite.

Ice Cored Moraine: Any morainic ridge marginal to an ice sheet or glacier, which encloses buried ice. This is formed either due to considerable accumulation of moraine in the glacier's terminal or lateral area, particularly in a state of ablation or due to the burial of ice avalanches or marginal snow banks by morainic debris. Ice cored moraines consist of several round topped curved ridges and are not normally sharp crested.

Terminal or End Moraine: Terminal moraines are formed at the end of the glacier or ice sheet where the front of the glacier remains stationary for a long period of time. These deposits are low ridge deposits arranged in crescentic manner.

Recessional or Stadial Moraine: When the glacier is receding during a short warm period with smaller ice volumes (stadial), the recession is of halting nature, i.e., it retreats and halts for a period of time.

Successive pauses in the position of glacier, while it is receding, leads to formation of successive terminal moraines known as recessional moraine.

Dead-ice Moraines: If the terminus of a glacier becomes stagnant, motion progressively decreases to zero, and surface, often obscuring the ice beneath a mantle of rock debris, is thick enough to allow forests id grow on when the ice completely melts away, a broad, irregular end moraine results, differing in form from other types of end moraines in its breadth and regularity. Such moraines are known as dead-ice moraines.

Push Moraines: Push moraines are formed by the bulldozing effect of an ice sheet advancing across the moraines formed during earlier glaciation.

Thule-Baffin Moraines: Thule-Baffin moraines refer to certain ice cored terminal moraines at the margins of cold ice sheets near Thule (Greenland) and in Baffin Island, Canada. Morainic debris produce a thick till cover along the glacier snout which protects it from ablation by direct insolation.

Ground Moraines: The excessive load carried by a glacier is deposited on its own bed or the base. This happens when the lower part of the ice becomes so heavily charged with particles that it cannot transport, and hence, it is deposited as ground moraine.

Washboard Moraines: Washboard moraines, (also known as De-geer moraines) so called because of its resemblance to washboard, is a type of cross valley moraine in which the relatively low ridges are closely spaced and are parallel to each other. It may be formed due to (i) seasonal pushing effect of a glacier front, (ii) concentration of ground moraine along the zones where thrust planes reach the base of the ice mass, and (iii) the squeezing out of subglacial till into marginal troughs along the ice front where it terminates in a proglacial lake.

Interlobate Moraines: Moraines deposited between two adjacent concurrently active ice lobes are described as interlobate moraines.

Ribbed or Rogen Moraines: Ribbed or Rogen moraines owe their origin to subglacial thrusting and, are therefore, allied to it. They consist of fairly regularly spaced ridges, usually arcuate in plan and concave upglacier, typically 10 m or more in height and up to 1 km long. They may form through preferential deposition where shear stresses, transverse to the direction of flow, are relieved and experience subsequent streamlining by the ice.

Temperamental Blocks

Erratic is a rock that has been transported by ice and deposited in an area of dissimilar rock type. Volcanic and metamorphic rocks make the most impressive erratics, since they are often distinctly coloured and their high resistance to erosion permits their continued existence despite considerable transportation. These are normally large, some have been left stranded in precarious positions and are known as perched blocks. Some erratics have come from sources 1000 km distant. If an erratic is so distinctive in character that its source can be unequivocally identified, it is known as indicator. Some indicator sources are so localised that their stones are spread over a fan-shaped area downstream, forming a boulder train or indicator fan.

Tillite: Tillite is a term used for clays and tills of earlier periods of wide spread glaciation, e.g., Carboniferous. These have compacted and lithified to form a tough sedimentary rock.

Drumlins: Drumlins are low, smoothly rounded, oval hills of glacial drift, generally of till having a steep (stoss) side facing the ice whose axis parallels glacier direction. Drumlins range from few hundred metres to 1 to 2 km in length and 10-60 metres in height. They resemble an inverted canoe shape or an inverted bowl of spoon. Drumlins exist in a wide range of sizes and shapes, most are elongate hills about 1 to 2 km long, 300 to 600 m wide and less than 50 m high. The ratio of their length to their width is generally about 2:3:5. The proximal portions of drumlins are usually steeper and higher than the distal portion which is thin and narrow in the down ice direction. Drumlins seldom occur in isolation.

Drumlins usually occur in groups termed a field or a swarm popularly known as baskets of egg topography. The basket of egg topography contains as many as 10,000 individual drumlins. Drumlin fields exist in Ireland, England, the Canadian Rockies and Prairies, Wisconsin, Michigan, upstate New York, western Washington and New England. Most drumlin fields occur in zones just behind end moraines with their long axes at right angles to the end moraines. Where the ice margin is lobate, the axes of drumlins diverge radically towards the curving arcs of the end moraines.

The material composing drumlins varies over such a wide range that drumlins could be said to consist of almost anything. Many are composed of till corresponding to the ground moraine between drumlins, but others may have a veneer of such till over a core of other material, such as, older till, glaciofluvial debris, or even bedrock. Such forms are cored drumlins. Drumlins have clearly much of the streamlining is due to deposition and how much to erosion is still a matter of debate.

It is speculated that drumlins form rear the edge of on ice sheet under conditions of unusually heavy debris load in the basal ice. The ice may be greatly slowed, but it must be moving enough to shape the deposit as it is laid down. Lodgement till, rich in fine particles and containing considerable water, expands under stress, becoming more mobile. This behaviour is thought to facilitate development of the streamlined form.

Drumlins may grade into narrow, elongate parallel ridges known as flutes.

Most drumlins show no internal structures while some exhibit a degree of stratification, particularly those, which contain a very heterogeneous collection of drift. There is no agreement on their mode of formation. Basically, there are two hypotheses. Erosional hypothesis suggests that the basal ice moulded a pre-existent landscape of glacial drift. Depositional hypothesis suggests that the basal till was deposited around a nucleus of rock or frozen drift which was moulded by ice movement. Classic examples can be seen in Northern Island in New Brunswick near Boston, Massachusetts, in southern Ontario and central New York (USA).

Till Deposits of Stagnant Ice

Glacier Table: Glacier tables are isolated slabs of large blocks supported by a pedestal of ice on the surface of a glacier. Rocks or patches of debris when sufficiently heated by the Sun may melt the underlying ice but this is not true of large blocks. Large blocks protect the underlying ice from melting by direct insolation while the surrounding glacier ice is lowered by excessive ablation, leaving the blocks perched on columns. Eventually, the blocks fall.

Glaciofluvial or Stratified Deposits: Sediment transported by meltwater can be deposited either in contact with the glacier or beyond the ice front in the proglacial environment. Most of the deposits that take place because of ice contact gives rise to features formed in channels while some may be constructed either subglacially or marginally in lakes or on land.

Dominant Sediment	*Environment*	*General Form*	*Relationship to Ice*	*Genetic Term*
Ice-contact deposits		Ridge	Marginal, sub-glacial, englacial, supraglacial	Esker
Sand and gravel	Fluvial			
				Kame
		Mound		Kame complex
		Spread with depressions	Marginal	Kettled sandur
Proglacial deposits	Fluvial	Spread	Proglacial	Sandur
Sand and gravel				
Silt and clay				Lake plain
Sand and gravel	Lacustrine	Terraces, ridges	Proglacial/marginal	
Clay sand		Beach Terrace		Kame delta
Silt and clay		Spread		Raised mud flat
Sand and gravel	Marine	Terraces, ridges		Raised beach
Clay, sand and gravel		Terrace		Raised delta
Ice contact deposits				

When contact with an ice mass has controlled or strongly influenced the shape, configuration and character of any glacial depositional feature, it is classed as ice contact feature. Ice contact features on land are kames, eskers and kettle holes.

Kame Family: Kames include a whole family of stagnating ice forms, many of which have more specific names. The term kame, by itself, is generally used for irregular mounds of sand, gravel and till that have accumulated in depressions or cavities in or on stagnating glacier and dumped anywhere on the land surface when the ice melts.

The surface of the ice has numerous basins and depressions, which have formed, because of thawing action. These depressions get filled by sediments carried by surfacial, melt waters emanating— from the top of the glacier. As the ice melts the material, that formerly filled depressions on top of the— glacier, is dropped and makes small hills. They are known by different names—kame, kames, kame— complex and kame terrace.

Kame terraces are formed by the lateral and frontal accumulations of fluvio-glacial deposits along the ice margin. Kame complexes result from the letting down of numerous sediment-filled supraglacial depressions and cavities in stagnant ice on to the subglacial surface.

Kettles: Kettles are depressions formed by the melting of buried ice. The size of the kettle depends on the size of the buried blocks of ice and may range from a few metres to more than a kilometre in diameter.

Crevasse Filling: One special type of kame is crevasse filling formed by the dumping of materials that filled crevasses. It is ridgelike rather than a hillock.

Eskers: Eskers, also known as osser or oss, are sinuous, sometimes discontinuous ridges formed of sand, gravel and boulders and range up to 200 m in height, 3 km in width and 100 km or more in length. Eskers are common in areas of continental glaciation where ice can stagnate and not in valley glaciers where ice rarely stagnates. Eskers are typically composed of eater laid, coarse grained, stratified, sorted sand and gravel, usually pebble/cobble-sized material but with occasional boulders. Bedding is almost always present but may be irregular and cross bedding is common.

Although, some eskers form by accumulation of sediment in supraglacier channels, in crevasses, in linear zones between stagnant iceblocks, or in narrow embayments at glacier margins, most eskers are believed to have been deposited in ice walled tunnels at the base of glaciers on the basis of the nature of the water washed sediments and their morphology. Water in supraglacial channels, moulins and englacial cavities collects in subglacial tunnels. The tunnels remain open as long as inflow of the enclosing ice is balanced by melting of the ice making up the tunnel walls and by the hydrostatic pressure of the water filling the tunnels.

Eskers may form in either confined or unconfined systems. Unconfined drainage may occur in supraglacial cavities or in some subglacial tunnels. Confined drainage occurs in subglacial tunnels that are hydrostatically closed and the water may flow either uphill or downhill under hydrostatic pressure. Good examples of eskers can be found in Ireland, Scotland, Finland, Sweden and in Maine, USA.

Some eskers are beaded, i.e., they pinch and swell in regular pattern. The beads in the esker represent deltas formed at the ice front during halts in its recession. There may have been an outwash fan or cone at the ice front. As the ice retreated, the cone was added from behind so that the ice left behind a more or less continuous gravel ridge. If the ice front halted at regular intervals of time, these would be represented by local broadening of the esker where the stream issuing from the ice had opportunities to build up bigger fans. Thus, beaded eskers were formed.

	Eskers and Terminal Moraines	*Eskers and Crevasse Filling*
1.	Terminal moraines are large unstratified forms, eskers are more regularly bedded.	Crevasse filling is usually smaller than eskers.
2.	Terminal moraines at right angles to ice movement direction. Eskers are approximately parallel to it.	Crevasses filling may extend in any eskers are direction, parallel to the direction of ice move.

Proglacial Deposits: Proglacial deposition is dominant in the proglacial zone both because of the decreased capacity for transport of meltwater streams, once they emerge from the ice front and the common presence of margin lakes in this environment. The rapid dumping of large quantities of coarse debris immediately beyond the glacier margin and the associated shifting of stream channels (because of braiding) produces an extensive depositional plain known as sandur (plural sandar). Sandar are broadly analogous to alluvial fans with characteristic entrenchment by the main feeder channels but they experience much more drastic seasonal fluctuations in discharge.

Valley sandar form in the laterally confined environment of glacial troughs, whereas plain sandar or outwash plains can develop along the margins of ice sheets where braided rivers from numerous outlets along the ice front coalesce. Streams issuing from the front of the glacier that are braided and frequently shifting as they are heavily laden with debris, derived from ice. Thus, even a small decrease in gradient leads to deposition. These plains, called outwash plains, have sand and coarse gravel near the source, with sand further away and finally, the clay.

The surface outwash plain is commonly interrupted by kettles, especially near the terminus of the glacier, giving the outwash plain a pitted appearance. Usually, large kettles in an outwash plain occur when melting of stagnant ice detaches large blocks from the glacier, which are then surrounded by outwash sediments. Melting of the enclosed blocks can leave kettles that may be more than 1 m in diameter where outwash sediments are deposited over stagnant ice, wholesale collapse of the outwash plain may occur when the buried ice melts.

During glacier retreat, a stream's sediment load is reduced and the under loaded stream cuts down into its outwash deposits to produce outwash terraces. Series of terraces are common in valleys that have experienced repeated glaciations. Generally, each major terrace can be traced upstream to an.end moraine or former ice limit.

Deposits of Glacialacustrine

Glaciolacustrine deposits are deposits laid down in contact with ice in a lake.

When the ice terminates in lakes, it leads to formation of deltas (homopycnal flow). The fine material deposited in the lake leads to formation of varves (Swedish; 'varv' -lap in race). These are thin laminar beds of sediment divided into a thicker, lower, lighter coloured band of sand grading upwards into a thinner upper darker coloured band of silt. It is formed in a proglacial lake (formed as a result of damming by ice sheets as they advance into ice-free area) by the melting of glaciers. The summer melt season is represented by the lower coarser sediment (when the run-off is high). During the winter, when the lake is frozen over, no new sediment enters the lake and the fine mud settles out of suspension to form the thin dark layers. Since the two bands which comprise the varve are equivalent to one year sedimentary accumulation, the age of the deposit can be calculated by counting the varves.

Permafrost Condition

Permafrost is a condition existing below the ground surface irrespective of its texture, water content or geological character in which the temperature has remained below 0°C continuously for more than two years and if pore water is present in the material, a sufficiently high percentage is frozen and this helps to bind the mineral and organic particles. The term permafrost was first introduced by S.W. Muller in 1947 to describe permanently frozen soil but now its meaning has widened. The primary control of permafrost is climate. An equilibrium is established when the perennially frozen layer attains a depth determined by the atmospheric temperature and the geothermal flux. Since on an average, the flow of internal heat is permafrost table mean annual ground temperature maximum monthly ~ mean temperature less than 0.1 per cent of that from the Sun, the tendency towards seasonal equalisation of the atmospheric and surface temperatures level to zero amplitude is strong. Below the surface, the amplitude of the minimum monthly mean temperature seasonal fluctuation gradually diminishes until at about 15 m, where the temperature of the ground remains stable throughout the year.

This is known as the level of zero annual amplitude. The temperature at this point is below freezing point about -16°C. In reality, a close approximation exists between level of annual amplitude and the mean annual air temperature. With decreasing depth, the temperature rises steadily at a rate of about 1°C per 30-40 m until freezing point is reached and where permafrost ceases.

Climate (mean annual temperature) largely determines the distributional pattern of permafrost region. About 15 million sq km area in the Northern Hemisphere and the whole of 132 million sq km area in Antarctica is permafrost. In the Northern Hemisphere 40-50 per cent of the land surface in Canada, 86 per cent area in Alaska and 47 per cent of area in the erstwhile USSR is underlain by permafrost.

The upper limit of permafrost is an irregular surface known as permafrost table and the lower limit at which unfrozen ground appears is the talik. Permafrost can be continuous, discontinuous and sporadic.

In areas of continuous permafrost, the mean annual temperature, below the level of seasonal temperature fluctuations (10-15 m), is less than -5°C. Permafrost reaches a depth of 150 m or more but is absent beneath large water bodies. In the CIS the continuous permafrost occurs where the mean annual temperature is -8°C or less. The mean annual temperature in discontinuous permafrost varies between -5°C and -1°C below 10-15 m depth of the surface. Discontinuous permafrost extends to a depth of 30-150 m but is marked by those areas within.

Permafrost is of two kinds depending on whether or not there is excess ice in the ground. Excess ice forms when ice builds up in the ground and physically separates soil particles. This is called as segregated ice. When ice builds in pores in the spaces between soil particles it is known as pore ice.

Periglacial Process

There are two main ways in which ice in periglacial areas can modify the landscape. The first is through landforms associated with permafrost and the second is through landforms associated with freezing and thawing on an annual or short term basis. In these periglacial and active layer, certain geomorphological processes such as weathering (mechanical and chemical), mass wasting, and erosion, and transportation takes place.

Congelifraction: Congelifraction (Latin congelareto freeze; to break) is a geomorphological term that was introduced by K. Bryan in 1946 to denote the mechanical weathering of rocks by the freezing of interstitial water, leading to expansion, fracturing and disintegration of rocks. This alternate diurnal freeze-thaw cycle is confined only to the active layer. The effectiveness of frost splitting depends on the size of the pores. If the pores are very small the water may become supercooled but remains in liquid state. Repeated freezing and thawing of clay minerals can disaggregate mineral grains as small as 1 mm in diameter.

Congeliturbation: Congeliturbation includes frost heaving and frost thrusting.

Frost Heaving: Frost heaving is the process of lifting of mineral, soil and surface particles into surface hillocks due to pressures caused by the freezing of groundwater and ice formation in the soil. As ice nuclei form in the soil, free water migrates towards these nuclei which grow as veins or lenses of ice while adjacent sediments are dewatered. The grain size and pore size of sediments determine the rate of ice crystal segregation and water movement. Frost heave occurs not only in the autumn freeze back period but also during winter when ground temperature is below 0°C.

Frost Thrusting: Frost thrusting is the lateral movement of soil and other ground surface debris caused by the freezing of groundwater.

Chemical Weathering: The role of chemical weathering is negligible in the periglacial areas. Nevertheless, processes such as hydration, hydrolysis and oxidation occur when aerated water contacts minerals. However, at low temperature and with few microorganisms, these chemical processes proceed only very slowly. Where sea salts collect by evaporation under low atmospheric humidity some hydration might be expected.

Solifluction: Solifluction literally means soil flowage and occurs widely in periglacial regions. When the permafrost begins to thaw from the surface downward in spring and early summer, the upper zone of the soil becomes saturated and starts flowing down even on the most gentle slopes. It may seem that solifluction is caused by lubrication of soil particles but by causing the soil particles to lose their shear strength, owing to loss of friction and cohesion, lead to soil flowage.

Gelifluction: Gelifluction is one type of solifluction that is associated with frozen ground covering both seasonal freezing in addition to permafrost. Suitable conditions for gelifluction occurs in areas where downward percolation of water through the soil is limited by the permafrost table and where the melt of segregated ice lenses provide excess water which reduces internal friction and cohesion in the soil.

Congelifluction: The term congelifluction is applied to the progressive flow and movement of rock and Earth down slope, as a result of freezing and thawing of ice. Its meaning differs from gelifluction, which also includes seasonally frozen ground.

Along with these processes of mass movement, a variety of other mass movements are common features of periglacial morphogenesis. Snow and rock avalanches are frequent especially during spring time. Mud flows and debris flows break out of gelifluction lobes and sheets, and are an extension of the gelifluction process. Creep is common on slopes where gravity leads to sliding of the materials that have been frost heaved.

Nivation: Nivation refers to the localised erosion of a hillside by frost action, mass wasting and the sheet flow or rill work of meltwater at the edges of and beneath lingering snow patches creating hollows. Frost weathering is accentuated around their margins while mass movements continue to erode their floors. The degree of nivation depends in part on the presence or absence of permafrost in the underlying material, for in permafrost free areas, snow patch erosion can continue in the snow free areas as well. The main effect of nivation is to produce nivation hollows, which, as they grow in depth, trap more snow and thereby enhance the process of deepening. Permafrost promotes widening rather than deepening of the hollows.

Fluvial Processes: Rivers in the periglacial zone have extremely complex hydraulic geometries and a very different river regime as compared to rivers in temperate latitudes. High gradient mountain streams and large rivers on plains continue to flow under ice cover throughout winter whereas all but the largest rivers are completely frozen. However, the discharge is greatly reduced because of the lack of overland flow in winter.

When the thaw arrives in spring, a huge volume of water is released causing widespread inundation, the headwater streams swell and burst their ice cover or flow over frozen river beds, progressively developing a flood wave that moves downstream. Braiding of the stream courses and frequent shift in stream is a common phenomenon because of violent fluctuations in discharge, coarse nature of the debris (resulting from accelerated movement on hillslopes), weak soil development and poor binding of the vegetation cover.

When large rivers flow Poleward from south to north in Arctic North America and Siberia, i.e., into areas of more severe climate the thaw first begins in the headwaters and progresses towards the mouth. This results in masses of floating ice being jammed against the frozen downstream reaches. Water is ponded back and forced to spill across the flood plain while large ice flows continue to grind against the channel banks. For a brief period of time, the rivers move much larger loads and much coarser sediments than their mean annual discharge. When the blockage is finally released, a huge surge of water proceeds down the original channel capable of carrying with it immensely large boulders. The overall effect is the building up of alluvial cover of generally coarse grade, moderately well sorted but with occasional boulder beds and lenses of silt and clay where abandoned tributaries are being infilled.

The discharge of small periglacial streams, especially mountain streams follow a distinct daily cycle during the summer. The rivers reach peak discharge shortly after the hottest part of the day as overland runoff reaches the channels, the runoff nearly ceases before dawn after a freezing night. Under such conditions of daily cycle, sediment transport and deposition also undergo a daily cycle with alluvium deposited among shifting, braiding channels. The sediment load delivered to streams also fluctuates with the daily cycle of gelifluction and other forms of mass wasting. Except the mountain streams, most streams flow on aggraded beds. The mass wasting processes deliver sediments faster to the rivers than the rivers can remove them to the sea.

Wind Transport

Transportation by wind in the periglacial regions is quite important. The ability of wind to distribute silt and sand-sized particles under periglacial conditions is attributable to several factors—

1. Prevalence of strong winds which operate to full effect under sparse vegetation cover.
2. Abundant supply of suitable fine detritus furnished by broad alluvial flats on which the discharge at certain times of the year is so low as to allow the surface to dry out completely.
3. The growth of ground ice which tends to desiccate the uppermost layers of the regolith so that during the cold season a dusty surface is exposed to aeolian action, and
4. Blizzards which not only blow the protective blanket of snowfall but also the snow itself which carries much loose sediment with it.

Although the periglacial wind blown materials from periglacial regions vary considerably in size, nevertheless they can be broadly divided into two major classes—loess and aeolian sand.

Loess is normally an unstratified deposit composed mainly of silt sized quartz grains ranging in diameter from 0.015 and 0.05 mm. Most of the information on loess comes from the study of the equivalent Pleistocene deposits. Pleistocene loess locally attains thickness in excess of 50 m. The primary source of sediment for the loess is glacial outwash because deposits thicken and grow coarser towards the source and this is evident in the distribution of loess. Loess tends to be thickest immediately east of those rivers which during Pleistocene times carried a large amount of meltwater. Examples are afforded by Mississippi in the USA. Rhine and Danube in central Europe, and the Bug Don and Volga in the former USSR. However, not all loess comes from these sources; for example, much of the wind blown material in northern France appears to be of purely local origin, its composition varying according to the calcareous and non-calcareous nature of the underlying bed rock. Along the cool temperate oceanic margins, loess tends to be relatively sparse because the moist climate has prevented total desiccation of the surface; the relatively rapid colonisation of the outwash by plant communities may have been another factor.

Aeolian sand in periglacial regions may assume two forms; one in the form of dunes and the other as a relatively smooth plain—the sand sheets. Dune formation implies an abundant supply of sand provided by heavily laden meltwater streams. Dunes are common feature on downwind banks of rivers. Sand sheets are less well sorted than dunes and sometimes stratified. They are widespread in low countries of northwestern Europe, known as coversands. These deposits originate during winter blizzards when both snow and fine sediments are blown by the wind. When the snow melts, the paniculate matter that is left behind slowly undergoes accretion. The resulting deposits are sometimes called niveo-aeolian sands.

Characteristics of Earth

Many of the features found on the surface of the Earth, whether on continents or in the oceans are not stable and changeless. There is a definite change, although it may be slow. These changes are continuously occurring both on the surface and beneath the surface. Normally, these changes are very slow, so slow that they are imperceptible but occasionally, particularly when volcanic eruption takes place or when landslides occur, we can witness sudden changes taking place.

The surface relief or topography results basically from the action of two types of forces—

a. Endogenic, Endogenetic or Hypogene forces, which give rise to land upliftment, subsidence, folding, fracturing and volcanic eruptions. In general, these forces are responsible for the major structural units of the Earth's surface, for example, mountains, plateaus and plains.

b. Exogenic or Exogenetic or Epigene forces give rise to destruction carving, moulding and smoothening of the major relief features. They produce the intricate details of the surface topography, for example, waterfalls, hills, valleys, spurs, dunes, caves, stacks, etc.

The various types of Earth movements result from the work of endogenetic forces.

System of Material Distribution

The theory of isostasy postulates a system for the distribution of material in the earth's crust which conforms to and explains the observed gravity values.

The theory was developed from gravity surveys in the mountains of India about 1850. A series of measurements were made at different elevations across the mountain range. These observed values were then reduced to what they would have been if the mountains were levelled off and the observation made at sea level. These values were then compared with the theoretical value for gravity at that latitude. The results show that the actual force of gravity over the mountain range is considerably less than it should be theoretically. In fact the difference is so great in some cases that the two values are more nearly the same if the Bouguer correction (which removes the attraction of the rock material between sea level and the point of observation) is not made. This discrepancy between the observed and theoretical values can be explained only in terms of the distribution of rock densities below sea level.

Put simply, the concept of isostasy means that a state of equilibrium or balance exists in the Earth's crust. Moho (the boundary between the crust and mantle) is found deepest beneath the highest mountain ranges and shallower beneath regions of lower surface elevations. For example, the Moho is 80 km deep beneath large mountain ranges, 40 km beneath the continental regions and only 6 km beneath the ocean bottom. This means that the lithosphere is 'floating' on the asthenosphere with more massive portions reaching deeper to obtain needed buoyancy.

It can be understood thus—a massive block of ice will extend deeper into the water than a small ice cube. Its top side will be projecting into the air proportionately. If several blocks of wood of the same cross section but of differing height are floating in a tank of water, each of the different blocks raises itself above the water level by an amount, which is proportional to its length. A series of such blocks are in a state of hydrostatic equilibrium. Likewise differences in density may also be responsible for such hydrostatic balance. Lighter density materials have a tendency to rise high. Thus mountains which are composed of lighter materials than the surrounding lowlands are high.

Now if an inch was sawn off one of the blocks, that particular block would adjust itself by rising in the water and a part of the block previously submerged would adjust itself above the water level while the block would not sink as much as it originally did. Loss of material and thus, weight by denudation of a landmass will upset the equilibrium of the landmass and cause it to rise until equilibrium is reached. If any additional weight is added to the landmass, such as a pile of volcanic material on the oceanic crust, it can make the crust subside, or accumulation of ice will also tend to depress the landmass. This process must involve some compensatory lateral flow of material at depth in the mantle, i.e., there must be lateral flow from beneath the sinking column to under the rising column.

Principles of Density Distribution

Two systems of density distribution were set forth by Airy and Pratt in 1855. These two theories have been modified very little since their presentation, and their relative merits are still debated. A third such type of theory was put forward by Heiskanen.

Airy Theory: Airy postulated that there is a change in the density of rocks at depth in the earth's interior and that the upper lighter material floats on the more dense part, which behaves like a fluid. The depth of this change varies from place to place. Under mountains the depth is greater than under oceans or plains. In other words the mountains have roots of lightweight materials which extends down into the lower, more dense mantle rocks. The lower density (mass) of the roots causes the observed gravity values measured in the mountain ranges to be lower than predicted theoretically.

Pratt Theory: Pratt's theory differs from Airy's in that Pratt assumed that the boundary between the upper light material and the lower dense rocks is at a uniform depth, called the depth of compensation. He further postulated that there are variations in the density of the lighter layer, which are related to the elevation of the surface. Lighter material lies under mountains, and heavier material under oceans. The weight of columns of rock extending from the surface to the depth of compensation in different parts of the earth is thus the same. Gravitational studies offer no means of determining which of these theories is more nearly correct. It is possible that neither is absolutely correct, but evidence from other geological studies suggest that Pratt's theory comes closer to fitting all the facts we have. Seismic studies indicate that more dense material underlies ocean basins than continents. These observations support Pratt's assumption of a connection between density and elevation.

Heiskanen's Theory: Airy postulated that columns of rock are nearly the same in density, floating like blocks in a fluid, Pratt suggested that different segments of the earth are of different densities, but that the differences are compensated at a certain depth. A third hypothesis has been formulated by Heiskanen (1933). He combines the assumptions of both Airy and Pratt. It has been observed that rocks at sea level are more dense on the average than those at higher elevations (2.76 grams per sq cm at sea level down to 2.70 gm/sq cm at elevations in high mountains).

He assumes that this change continues downward, tending to make deeper rocks more dense than shallower ones in all sections of the earth's crust, In addition, different sections are thought to have different densities and different lengths. Heiskanen's theory has the advantages to being based on actual knowledge of density variations that can be obtained by direct measurement, and when it is applied to the observations we now have on the gravitational field it yields very low anomalies.

Modification of Phase

The alternative explanation of vertical movements is phase change. Some materials, when subjected to considerable pressure, change their atomic and crystal structures and compact themselves into a smaller space, i.e., they undergo phase change, Olivine, which forms a large proportion of mantle material, behaves in this way. Thus, at a certain level in the mantle, where critical pressure is reached, a phase change occurs and a phase change boundary can be drawn in. If erosion occurs, the pressure on an area of the Earth's crust is decreased, and the phase boundary line moves downwards in position (i.e., the critical pressure for change is not attained until a greater depth is reached). This means that some of the mantle material is changed from a high density to a low density form and consequently increases in volume. This, in turn, causes an uplift of rock column above it. Conversely, loading by ice, sedimentation, piling of volcanic material at the Earth's surface produce an upward movement of the phase change boundary, a compaction of some of the mantle material and a resultant sinking of the rock column above.

Denudation and Isostasy

Since the rate of denudation is functionally related to relief and the latter is reduced through time, it is possible to calculate, on the basis of the above relationship, that after 11 m.y. landmass would be reduced to 10 per cent of its original relief and after 22 m.y. it should be reduced to 1 per cent, as the rate of denudation declines both with time and relief.

Denudation rates, however, refer only to material stripped from the continental surfaces. In considering destruction of relief we must also take into account isostatic rebound, as the gradually thinning continent becomes more buoyant on the underlying mantle. Isostatic rebound can be calculated thus:

$h = Br/A$

Where, h = isostatic compensation,

B = specific gravity of surface rocks removed.

A = specific gravity of material replacing at depth, and

r = thickness of surface layer removed.

Assuming $A = 3.4$ and $B = 3.6$ then $h = 0.76\ r$.

Assuming then, that isostatic rebound, in response to erosional losses, is a continuous and widespread process, then three-quarters of the relief removed within a given period of time is replaced. Accordingly, the real length of time taken for the destruction of the relief of a land mass will be longer, and it is calculated that, with isostatic compensation, 18.5 m.y. will be required for reduction of 10 per cent of initial relief and 37 m.y. of 1 per cent reduction.

Eustatic Movements: Eustatic Movement (Greek: 'ew'-well, *'statis'* -standing) involves the world wide movement of sea level resulting from changes in the total volume of liquid sea water and the capacity of the ocean basins. For example, rapid convection in the mantle would arch up the oceanic ridge and displace water from the ocean basins to the continents as the capacity of the ocean basin decreases.

If the convection rate decreased, the oceanic ridges subside and the seas would recede with a fall in sea level. As the plates move, the sea level rises and falls slightly and thus causes expansion and contraction of shallow seas over the cratons. If the continental mass gets arched up, there is a general withdrawal of sea.

Sudden Movements: Sudden movements include Earthquakes and Igneous movement.

Features of an Earthquake

Earthquakes are vibrations or oscillations on the surface of the Earth caused by the rupture and sudden movement of rocks that have been strained beyond their elastic limit. The upper part of the Earth's crust and upper mantle (lithosphere) are very strong and brittle. But when this rock is subjected to deformation, the rock actually bends elastically.

The rock is able to withstand only slight stress with only slight bending or strain. Continued stress leads to the building up of strain up till a point. This is the elastic limit of the rock. If the rock is strained beyond this limit, a fault is formed and the bent rock snaps up to regain its original shape releasing the stored energy in the form of rebounding and violent vibrations. These vibrations shake the ground even hundreds of kilometres away. The shaking is strongest along the fault.

When the vibrations are felt in the bedrock and ground, they are called shocks. The point within the Earth where Earthquakes are generated is called focus. The point on the Earth's surface directly above the focus is called the epicentre. Earthquake focus can be shallow (up to 70 km depth) known as shallow focus earthquake, between 70 km and 300 km it is known as intermediate focus earthquake and the focus below 300 km up to 700 km is deep focus earthquake.

Effect of Earthquakes: The dangers of earthquakes are profound and the damage they do is catastrophic. Their effects are of two types—Primary and Secondary. Primary effect cause damages directly and includes ground motion and faulting; secondary effects cause damages indirectly as a result of process set in motion by the Earthquake.

Primary effects

1. Ground motion results from the movement of seismic waves, especially surface waves, through surface rock layers and regolith. The motions can damage and sometimes completely destroy buildings.

Where a fault breaks, the ground surface buildings can be split, roads disrupted and any feature that crosses the fault can be broken apart. The 1906 San Francisco Earthquake led to the visible displacement along the San Andreas Fault. Horizontal displacement occurred over a distance of about 400 km and offset roads, fences and buildings by as much as 7 m.

Secondary effect

1. In regions of hills and steep slopes, Earthquake vibrations may cause regolith to slip, cliffs to collapse, and other rapid mass-wasting movement to start. This is particularly true in Alaska, parts of southern California, China and hilly places such as Iran and Turkey.
2. The sudden shaking and disturbance of water saturated sediments and regolith can turn seemingly solid material to a liquid like mass, such as quick sand. This is called liquefaction and it was one of the major causes of damage during the earthquakes that destroyed much of Anchorage, Alaska, 1064.
3. Seismic sea waves occur following violent movement of the seafloor. Also called Tsunamis, they occur particularly in the Pacific Ocean.
4. Earthquakes can cause fire by breaking gas lines and snapping electric wires.

Important Terms

Anticlines	Exogenic	Plunge
Axial plane	Footwall	Plunging fold
Box folds	Hanging Wail	Recumbent fold
Chevron fold	Horst	Reverse fault
Cleavages	Hypogene	Rift valleys
Diastrophic Movement	Kink bands	Rotational fault
Earthquakes	Linear structures	Step faulting
Endogenetic Movement	Monocline	Strike-slip faults
Endogenic	Nappe	Symmetrical
Epeirogenic Movement	Normal Faults	Tectonic Movement
Epigene	Oblique-slip fault	Transcurrent fault
Eustatic Movement	Overturned fold	Wrench faults
Exogenetic	Parallel folds	
	Isoclinal folds	

Earth's Movement

Endogenetic Movements: Endogenetic Movements (Greek: *'endo'*-within, *'genera* '-origin) fall into two major categories—Diastrophic Movements and Sudden Movements.

Diastrophic Movements: (Greek: *'diastropos'*-turned, twisted, distorted) are those which result directly or indirectly, in relative or absolute changes of position, level or altitude of the rocks forming the Earth's crust. These diastrophic movements lead to the formation of primary landforms. Primary landforms vary in size from continents to miniature Earthquake fault scarps. There are three major classes of diastrophic movements, all of which are interrelated.

1. Tectonic Movements.
2. Isostatic Movements.
3. Eustatic Movements.

Tectonic Movements: Tectonic movements (Greek: *'tekton* '-a builder) includes epeirogenic and orogenic movements.

Epeirogenic Movements: Epeirogenic movements (Greek: *'epeiros'*-land, continent) relate to the behaviour of continental platform or stable blocks involving broad gentle upwarping of relatively large crustal areas. These are vertical movements, caused by radial forces and are characterised by large scale upliftment, subsidence or submergence and emergence of land areas. The movements involved are so slow and widespread that no obvious fracturing or folding is produced in the rock.

Every continent including Antarctica offers unmistakable evidence of both downward and upward movements since Precambrian times. The proof of such movements is found in marine sedimentary rocks of Palaeozoic, Mesozoic and Cenozoic age, which lie within continents upon well-eroded older rocks.

Much of the northern coast of the Gulf of Mexico has been subsiding very slowly for many centuries. Numerous islands off the coast of British Columbia are believed to have been, till recently, part of the mainland since they have submerged there. The seaways between the islands represent lowlands now submerged. Dredging in the North Sea, between England and Norway, which forms part of stable region of NW Europe, and has revealed stumps and roots of trees which once grew on land and that is now beneath the sea.

Movements of greater magnitude are demonstrated by the sedimentary successions built upon continental shelves and within the continents themselves. A vast majestic depression is recorded in the transgressive series of sediments deposited in spreading epicontinental seas as illustrated by Cenomanian transgression which drowned enormous low lying areas towards the end of Cretaceous period. On the continental parts of stable blocks, various types of sediments may accumulate. They include *arkoses* formed by the decay of uplifted mountain masses, aeolian and lacustrine deposits, coal, evaporites and tillites. The Karroo sediments of South Africa provide an illustration of the sequence formed by long continued deposition in the interior of relatively stable continental mass. Where no sedimentary succession have been formed, records of epeirogenic movement may be preserved by features of the surface topography. Lake Victoria and Lake Kyogo in East Africa lie within a shallow basin of recent origin and their sinuous outlines indicate the drowning of land.

True epeirogenic movements involving cratons are possibly due to mantle plume or due to the collision of two continental plates. Local circular elevations can occur as fast as 8 mm per year and these are associated with mantle convection plumes. West of Yellowstone National Park, an area of 8000 sq km is rising at a rate of 3-5 mm/year. During the period 1955-73 the central Adirondacks in New York State rose 40 mm whereas the northern margin subsided 50 mm. The Rhine Massif has been elevated 300 mm since the Pliocene (probably episodically). Contemporary uplift measurement rates of the Massif are 0.35 mm per year. A very characteristic behaviour of cratons is the long continued rise of localised plateaus. The Colorado Plateau seems to have risen about 550 mm in the Late Cenozoic at an average rate of about 0.1 mm per year and the uplift of Deccan plateau of India, during the Tertiary and Quaternary continues at present, at a rate of 0.36 mm per year.

Orogenic Movements: Orogenic movements (Greek: 'oros'-mountain) relate to behaviour of plate margins and involve intense folding, thrusting, faulting and uplift of narrow belts. They are caused by tangential forces acting on the surface of the Earth. These tangential forces belong to two categories—compressional forces and tensional forces.

Compression Force

Folding: A fold or buckle, in any pre-existing structure in a rock, is a result of deformation. Folds are best displayed by structures that were formerly approximately planar such as layering or bedding in sedimentary and igneous rocks, or foliation, schistosity and cleavage in metamorphic rocks.

If deformation is slight, the nature of the folding is easily described but it may be very complex and a geologist may have to employ careful mapping and measurement of structures in order to work out the nature, extent and age of the folds. Different terms define the various components of folds: The sides of folds are called as the limbs and the median line between the limbs, along the crest of an anticline or the trough of a syncline, is the axis of the fold. A fold with an inclined axis is called a plunging fold. The angle between a fold axis and the horizontal is the plunge of the fold. An imaginary plane that divides a fold as symmetrically as possible and that passes through its axis, is the axial plane.

In broad, open folding, the beds will retain their natural order of deposition, in both the upfolds called anticlines.

A syncline is formed when the strata are bent downwards. These structures in which the stratigraphy is unknown may be called antiforms and forms.

If the axial plane bisects the fold, the fold is said to be symmetrical or upright. If the axial plane has a dip, the fold is described as inclined or asymmetrical.

A sequence of fold structures may be formulated which are related to the degree of compressional forces involved.

Monocline, the simplest of folds, is a one-limbed flexure on both sides of which the strata are either horizontal or dip uniformly at low angles.

The angle between the fold limbs, as seen in profile, is a measure of the tightness of the fold. Folds with narrow hinge zone are described as angular, while if the hinge zone is broad they are described as rounded.

Folds with straight limbs and angular hinges may be described as chevron folds, if the limbs are symmetrical, and kink bands, if the limbs are markedly asymmetrical. Conjugate folds are apparently related folds with converging axial surface. Kink bands may form conjugate sets while conjugate folds with rounded hinges are box folds.

Parallel folds are folds in which the profiles of fold surface lie on arcs of circles. Similar folds are folds in which the thickness of layers is greater in the hinge zone than on the limbs.

In isoclinal folds, both limbs are essentially parallel, regardless whether the fold is upright, overturned or recumbent. An isoclinal fold results from the continued lateral compression upon an overfold crowding it upon the adjacent overfold. Here, both the limbs dip at equal angles in the same direction.

An overturned fold is one in which the strata in one limb have been folded beyond the vertical. Both limbs dip in the same direction, though not necessarily at the same angle.

A recumbent fold is literally a fold lying down resulting from the continuation of pressure, the axial plane and both limbs of the fold are roughly horizontal.

A nappe (French: *'nappe-a* cover) results when the pressure exerted upon a recumbent fold is sufficiently great to cause it to be torn from its roots and to be thrust forward. Nappes are basically large horizontal recumbent folds that have travelled for tens of kilometres along the thrust planes. They are well developed in the Alps and also in the Himalayas. A distinction is made between the autocthonous nappe, denoting rocks that are still in their place of formation and have not been displaced by thrusting, e.g., a folded sequence of rocks whose roots are still connected. Ailocthonous also called para-autocthonous, in contrast, denotes an isolated mass of rock displaced over a considerable distance from its original source by low angle thrusting. The plane that marks the boundary between two different types of deformation is called decollement. The rocks above are generally deformed, whereas those below may be unaffected. It is caused by the upper rock series sliding over the lower during folding.

Forces of Tension

Faulting: A fault is a fracture, which involves the displacement of blocks on either side of it, relative to one another. In faulting, there is movement of greater or smaller extent along the plane of fracture. Such a movement may be vertical or horizontal or a combination of both.

Most faulting occurs along features that are inclined. Faults can be described according to the attitude of the fault plane and relative displacement of the blocks on either side. The dip of the fault plane is the angle between the fault plane and the horizontal. A dip of less than 45° describes a low angle fault, while a high angle fault has a dip greater than 45°. Where the fault plane is not vertical, the side overlying the fault plane is called the hanging wall and the side underlying the fault plane is the footwall.

Faults are grouped according to—

1. The inclination of the surface along which fracture has occurred, and
2. The direction of relative movement of the rock on its two sides.

Normal Faults are caused by tensional forces that tend to pull the crust apart, and also by the forces tending to expand the crust by pushing it. Upward from below, movement on a normal fault is such that the hanging wall block moves down relative to the footwall block. In other words, normal faults are produced where tension in the Earths crust fractures rock and allows blocks to slip straight down. Normal faults with dip of less than 45° are sometimes known as lag faults or low angle (extension) faults.

Commonly two or more similarly trending normal faults enclose an upthrust. A down dropped block is a graben or a rift, if it is bounded by two normal faults or a half graben if it subsides along a single fault. An upthrust block is a horst.

Rift valleys are among the most dramatic fault features. The Mid Oceanic Ridge (MOR) system, running for about 64000 km, has its crest occupied by a rift valley. Perhaps the worlds most famous system of graben and half grabens is the Great Rift Valley of East Africa, which runs north-south all the way from Israel to Mozambique for a distance of more than 6000 km. The floor of the rift valley is the downthrown block called a graben.

Although none being as spectacular as the Great African Rift Valley, normal faults and graben are very common. The valley in which the Rhine River flows through western Europe follows a series of grabens. The Narmada and the Benue also flow in the rift valley. A parallel series of faults with a repeated downthrow in the same direction is termed step faulting.

A horst is the opposite of a graben and may create huge, high plateaus.

The mountains which are remnants of the Hercynian mountain building period, for example, Central Plateau of France, Vosges, Black Forest and Bohemian Massifs are all Block Mountains (in fact, they are relics of old fold mountains which have been denuded and shattered). The Sinai desert in the Middle East and the Ruwenzori Mountains in East Africa are all horst blocks. A horst and graben type of topography is found to be associated with Flinders Mountain and Torrens Valley in South Australia.

Reverse faults arise from compressive forces. Movement on a reverse fault is hanging-wall up relative to the footwall. Reverse fault movement pushes older rocks over younger ones thereby shortening and thickening the crust. Reverse faults are less steep and more varied in angle than normal faults.

A special class of reverse faults are called thrust faults (or thrusts). These are low angle reverse faults with dips less than 45°. Such faults, common in great mountain chains, are noteworthy because along some of them the hanging wall block has moved many kilometres over the footwall block. Thrust faults may be split into a kind of staircase of level sections called flats and step sections called ramp.

A normal or reverse fault on which the only component of movement lies in a vertical plane normal to the strike of the fault surface is the dip slip fault.

Patterns of faults develop according to the stress system involved, and may be parallel, en echelon, radial, concentric, or in two directions (intersecting).

Strike Slip faults (also called tear, transcurrent or wrench faults) are characterised by predominantly lateral (horizontal) movement. Movement of a strike slip fault is described by looking directly across the fault and by noting which way the block on the opposite side has moved. To an observer, standing on one fault block the movement of the other block is left lateral (sinistral) or right lateral (dextral). The sense of relative motion is the same regardless of the block on which the observer is standing. One form of tear fault is the transform fault. Transform faults are largely found on the ocean floor. They cut across and offset the rift zones of Mid Oceanic Ridges throughout the world. The most important surface expressions of transform faults are San Andreas Fault, California; Alpine Fault, New Zealand; the Philippines Fault and the Atacama Fault of Chile.

Oblique-Slip faults have similar magnitudes of both strike and dip slip components of displacement. It is a fault in which blocks of rock slip up and down and past each other diagonally. This occurs when a wrenching movement in the Earth's crust is combined with compression or tension causing blocks of rocks to slip diagonally past each other. If this happens on a massive scale, they are called transtension or transpression faults.

Rotational faults occur where the amount of displacement varies along the fault. The blocks on either side of the fault behave like rigid blocks rotating about an axis at right angles to the fault plane. A hinge fault occurs where a fault dies out and the end of the fault acts as a hinge. A scissor fault or pivot fault occurs where the sense of displacement changes between opposite sides of the axis about which the blocks rotate. There is an axis on the fault plane where no movement occurs; on either side of the axis, the displacement is in the opposite sense so that the action may be compared to opening a pair of scissors. Listric faults are spoon-shaped rotational- faults in which the hanging wall is rotated towards the fault in the same sense as the movement along the fault.

Movements of Isostasy

Isostatic (Greek: 'iso-equal 'statis-standing) movements involve vertical movements under the action of floatation displacement between rock layers of differing density and mobility to achieve balanced crustal columns of uniform mass above a level of compensation in which topographic elevation is inversely related to the underlying rock density. This isostatic movement is best represented by the readjustments, which followed the Pleistocene glaciation.

The formation of the great ice sheet over Scandinavia led to the land being depressed by the enormous weight of accumulated ice. When the ice eventually melted, the land surface began to rise in response to the gradual easing load. This uplift is illustrated by the occurrence of a series of raised beaches. The uplift is continuing and can be illustrated by changes, which have gone along the Bothnian coast of the Baltic, for example, the emergence of Alland Islands from the sea. More land is emerging above the sea and the already emerged islands are increasing in area. Because of this movement, it is possible that someday the islands may be united in a solid neck of land linking Finland to Sweden transforming the Gulf of Bothnia in a broad landlocked sea. The Finnish port of Vaasa has been in existence for several hundred of years but the modern port lies some six miles of the original harbour site, a fact which illustrated how the continuing uplift along this coast has rendered some of the old harbours quite useless.

Factors Affecting the Mechanisms of Glacial Abrasion

Fundamental Factors	*Comment*
Presence of debris in basal ice	Clean ice is unable to abrade solid rock. The rate of abrasion will increase with debris concentration up to the point where effective basal sliding is retarded.
Sliding of Basal ice	Ice frozen to bedrock cannot erode unless it already contains rock debris. The faster the rate of basal sliding, the more debris passes a given point per unit time and the faster the rate of abrasion.
Movement of debris towards glacier base	Unless particles at the base of a glacier are constantly renewed they become polished and less effective abrasive agents. Thinning of the basal ice by melting or divergent flow around obstacles brings fresh particles down to the rock-ice interface and increases abrasion.
Other factors affecting nature and rate of abrasion	
Ice thickness	The greater the thickness of overlying ice, the greater the vertical pressure exerted on particles on the glacier bed and the more effective is abrasion. This is the case up to a depth where friction between particles and the bed becomes so high that movement is significantly retarded and abrasion decrease.
Basal water pressure	The presence of water at the glacier base, especially when at high pressure, can reduce the effective pressure on particles on the bed, and thus abrasion rates by buoying up the glacier. However, sliding velocities may tend to increase because of the reduced friction.
Relative hardness of debris particles and bedrock	The most effective abrasion occurs when hard rock particles in the glacier base pass over a soft bedrock. If the debris particles are soft in comparison with the bedrock, the former are abraded and little bedrock erosion is accomplished.
Debris particle size and shape	Since particles embedded in ice exert a downward pressure proportional to their weight, large blocks should abrade more effectively than small particles. Moreover, angular debris will be a more efficient agent of abrasion than rounded particles.

6

Types of Energy

Earth's Seismic Zones

It may be observed that most of the earthquakes and volcanoes of the world are found in narrow, semi-continuous, but definite belts. These belts are for the most part confined to the plate boundaries. In fact, seismic belts have been used as a basis for the delineation of plates. Seismicity is limited to mid-oceanic ridges, trench arc systems, major fracture zones and the young folded mountain belts, and all these features are indicative of plate boundaries. On all plate boundaries, and especially on the mid-oceanic ridges shallow-focus earthquakes are found, but intermediate and deep-focus earthquakes, which have their focus deeper than 200 km, are mostly found within the oceanic trench zones. This is the area where the plate descends starting from the surface of the trench to depths of 300 to 400 km which may sometimes extend up to a depth of 700 km. The inclination of this zone is towards the continental land mass and the angle of inclination varies between 30° and 80°, though in most cases it is between 45° to 60°.

Such inclined seismic focus zones are found near the entire Pacific coast. They are called as Benioff zones *(Pacific Ring of Fire)* as the seismologist Hugo Benioff was the first to discover them.

As a general principle, it may be stated that the Benioff zone is related to the downward motion of the lithosphere.

On the basis of evidence from sea-floor spreading and palaeomagnetism, it has been now validated that the continents and ocean basins have never been stationary or fixed at their places rather these have always been mobile throughout the geological history of the earth and they are still moving in relation to each other. The scientists have discovered ample evidence to demonstrate the opening and closing of ocean basins. For example, the Mediterranean Sea, the Caspian Sea and the Aral Sea are the residuals of one very vast ocean *(Tethys Sea)* and the Pacific Ocean is continuously shrinking because of gradual subduction of American plate along its ridge.

Contrary to this, the Atlantic Ocean is continuously expanding for the last 200 million years. Red Sea has started to open (to expand). In the opinion of F.J. Vine, the Red Sea is spreading at the rate of 1 cm per year (total spreading 2 cm per year) since the past 3-4 million years. The Gulf of Aden is also widening at the rate of about 1 cm per year. Baja, the California peninsula, was previously united with the mainland of North America but later on broke away from the continent due to spreading of sea floor. Moreover, the Bay of Biscay is shrinking.

Criticism of the Theory: The geologists and geophysicist agree that the plate tectonics theory has been proved beyond doubt. There are, however, a few problems to which it has not been able to provide a satisfactory solution. The main criticisms of the theory are as under:

1. The length of spreading is far greater than the subduction zones. In other words, the rate of creation of new crust appears to be much greater than its rate of destruction.
2. Plate tectonics theory is unable to explain why subduction is limited to the Pacific coast while spreading is found in all the oceans.
3. There are evidence to show the movement of the same plate in opposite directions which is almost impossible.
4. The Benioff zone is not present equally in all the probable places. For example, the intermediate and deep focus earthquakes are absent in North America.

5. It cannot be said with confidence that all the plates behave like a unit, while some of the geologists have proposed an increase in the number of plates.
6. The theory of plate tectonics has failed to provide a fully satisfactory explanation for mountain building. There are several mountain ranges, such as the eastern highlands of Australia, Drakenburg mountains of South Africa and Sierra Delmar of Brazil which cannot be related to plate tectonics.

Despite all these omissions and commissions, plate tectonics is a revolutionary and comprehensive theory which scientifically explains the present distribution and arrangement of the continents and ocean basins. It also provides a satisfactory explanation of the distributional pattern of earthquakes and volcanoes in the world. It has also confirmed the theory of continental drift. The theory of plate tectonics has had the same effect on geology that the theory of evolution has had on biology. The theory has revolutionised our understanding of continental geology.

Geologists are now searching and recognising ancient plate boundaries—zones where old continents have been welded together—and how small continental fragments, ancient sea mounts, and island-arcs have been accredited to the continents. Plate motion has dominated tectonic processes on our planet for at least the past 2,500 million years, providing a fascinating record waiting to be interpreted. In brief, the theory of plate tectonics has produced a 20th century revolution in the earth sciences.

Spots that are Hot

A significant aspect of plate tectonics is the estimated 50-100 hot spots across the surface of the earth. These are individual sites of upwelling material from mantle. Hot spots occur beneath both oceanic and continental crusts and appear to be deeply anchored in the mantle, tending to remain fixed beneath migrating plates. Unlike convectional currents in the asthenosphere that principally drive plate motions, scientist now think hot spots are initiated by upwelling plumes that are rooted below the 670 km transition zone. Thus, the area of a plate that is above a hot spot is locally heated for the brief geological time it is there (a few hundred thousand or million years).

An example of an isolated hot spot is the one that has formed the Hawaiian-Emperor islands chain. The Pacific plate has moved across this hot upward erupting plume for almost 80 million years, with the resulting string of volcanic islands moving northward, away from the hot spot. Thus, the age of each island or sea mount in the chain increases northward from the island of Hawaii.

Iceland is also an example of an active hot spot sitting astride a mid-ocean ridge. It is also the best example of a segment of mid-ocean ridge rising above sea level. This hot spot has generated enough material to form Iceland and continues to cause eruptions from deep in the asthenosphere. As a result, Iceland is still growing in area and volume. The youngest rocks are near the centre of Iceland, with rock age increasing towards the eastern and western coast.

Volcanoes, Earthquakes and Plate Tectonics: Plate boundaries are the primary location of earthquake and volcanic activity, and the correlation of these phenomena is an important aspect of plate tectonics. The *'Ring of Fire'* surrounding the Pacific basin, named for the frequent incidence of volcanoes, is most evident. The subduction edge of the Pacific plate thrusts deep into the crust and upper mantle, producing molten material that makes its way back towards the surface, causing active volcanoes along the Pacific Rim. Such processes occur similarly at plate boundaries throughout the world.

Plate Tectonics and Mountain Building (Orogeny): The theory of Plate Tectonics is a comprehensive theory which has given scientific explanations to the drift of continents, occurrence of volcanoes, earthquakes, ocean trenches, relief features, isostatic equilibrium of the earth crust and the mountain building. A brief account of the plate tectonics (movement of the major and minor constructive and destructive plates) and upheaval of mountains has been presented in the following paras.

In general, the high standing folded and faulted mountains (other than the volcanic mountains) are elevated by one of the two basic tectonic processes: (i) *compression,* and (ii) *extension* (pulling apart).

The compressional tectonic activity *(squeezing together* or *crushing)* acts at converging plate boundaries which result in the building of folded mountains like the Rockies and Andes to the east of the destructive plate boundaries of the Pacific Ocean.

Contrary to this, the extensional tectonic activity *(pulling apart)* occurs where oceanic plates are separating like that the Mid-Atlantic Ridge. Such a situation also occur where a continental plate is undergoing break-up into fragments. The rift valley in East Africa and Jordan is an example of divergence or extension of plates.

Compressional Processes of Tectonic: Young folded mountain chains like the Alps, Himalayas, Rockies, Andes, etc., are made up of intensely deformed strata of marine origin. The strata are tightly compressed into wavelike structures, called *folds.* As compression continues, the folds are first overturned and then become *recumbent* as they are further overturned upon themselves. The folding activity is accompanied by *faulting.* In the process of faulting, slices of rock move over the underlying rock on fault surfaces of low inclination. These are *over thrust faults.* In the European Alps, thrust faults of this type were named *nappes* (from the French word meaning *'cover-sheet'* or *'table-cloth').*

As a result of arc-continent collision the Rockies and Appalachian mountains of North America came into existence, while collision of continent with continent resulted in the formation of Atlas Mountains of North Africa, Taurus Mountains of Turkey, Zagros Mountains of Iran, Hindu Kush of Afghanistan and the Himalayas of the subcontinent of India.

The European Alps were formed when the African plate collided with the Eurasian plate in the Mediterranean region. The Zagros Mountains resulted from the collision of the Arabian plate with the Eurasian plate. The Himalayan ranges represent the collision of the subcontinent of Indian portion of the Austral-Indian plate with the Eurasian plate.

Continent-continent collisions have occurred many times since the late Precambrian time. Several ancient *sutures* have been identified in the continental shields. The Ural Mountains, which divide Asia from Europe, are one of such sutures, formed near the end of the Palaeozoic era.

The divergent movement of plates or 'extension' is also responsible for orogeny and rift valley formations. Passive margins of plates are formed when a single plate of continental lithosphere is *rifted apart.* This process is called continental rupture.

In the case of passive continental margins, the crust fractures and moves along faults—upthrown blocks form mountains, while down dropped blocks form basins. In this process, eventually a long, narrow valley, called a *rift-valley,* appears. The widening crack in its centre is continually filled in with magma rising from the mantle below. The magma solidifies to form new crust in the floor of the rift valley. Crustal blocks on either side slip down along a succession of steep faults, creating a mountainous landscape. The Rift valley system of East Africa is a notable example of continental rupture. The Red Sea is also the result of continental rupture.

The theory of Plate Tectonics thus explains scientifically the origin of mountains and the processes of orogeny during the different eras and periods of geological history.

Over the long span of geologic time, the earth's surface is shaped and reshaped endlessly. Oceans open and close, mountains rise and fall. Rocks are folded, faulted and fractured. Arcs of volcanoes spew lava to build chains of lofty peaks. Deep trenches consume sediments that are carried beneath adjacent continents. These activities are by-products of the motions of lithospheric plates and are described by plate tectonics.

The main significance of plate tectonics to geographers is that it provides a grand scheme for understanding the nature and distribution of the largest and the most obvious features on the surface of the earth, its continents and ocean basins and their major relief features. However, large features, such as mountain ranges, are made up of many smaller features, such as individual peaks, gorges, spurs, terraces, etc. The formation of these landforms is often marked by powerful earthquakes, which influence the humanity and society.

Three Theories

There are three important assumptions which underline the theory of Plate Tectonics—

1. That sea floor spreading occurs, i.e., new oceanic crust is continuously generated at irregular line sources (active oceanic ridges).
2. That the Earth is of constant surface area, or if not, the area changes at a rate which is small by comparison with the rate of generation of new surface area by spreading.
3. That once formed new crust constitutes part of a rigid plate, which may or may not incorporate continental material.

Plate tectonics is the study of deformation within plates and of the interaction of plates around their margins.

Plates are broad rigid segments of lithosphere including the rigid upper part of upper mantle plus oceanic and continental crust that floats on the underlying asthenosphere. The lithosphere, the outermost layer and the only one directly accessible to us, is cold and rigid. Below it, is the asthenosphere, which is hot and capable of being slowly deformed. The asthenosphere is not liquid. The asthenosphere is a solid, but one that flows under stress. It is unlike ice, which is brittle in the form of ice cube, but like glacier quite plastic flowing down a mountain valley. The distinction between lithosphere and asthenosphere is based on mechanical properties (rigidity) and not on chemical composition which divides Earth into crust and mantle. The crust is the upper portion of the lithosphere. The lithosphere contains the topmost part of the-mantle. The asthenosphere almost entirely lies within the mantle. These plates are 100-150 km thick, have high flexural rigidity and are unable to deform except in response to very strong or prolonged forces. Plates range in size from 104 sq.km to 108 sq.km.

Distribution of Plates

There are seven major plates on the Earth's surface today. These are the Eurasian, Antarctica, North American, South American, Pacific, African and Indian Plate. Intermediate sized (106-107 sq km) plates are the China, Philippines, Arabian, Iran, Nazca, Cocos, Caribbean and Scotia plates. There are more than 20 plates with areas of 104-106 sq km and many of these are not precisely defined.

The Pacific plate is the largest plate composed almost entirely of oceanic crust and covers about 1/5th of the Earth's surface. The other large plates contain both continental crust and oceanic crust but none of them are entirely composed of continental crust.

Individual plates are not permanent features. They are in constant motion and continually change in size and shape. Plate can change its shape and size by splitting along new lines or by welding itself to another plate.

Plates are constantly in motion with respect to each other and to Earth's axis of rotation.

Movement of Plate Regarding Earth's Axis of Rotation: Because all the plates on the Earth are moving the motion of one plate is described relative to that of another, which is arbitrarily fixed. Since all the plates are moving on a sphere, it indicates that the motion of one plate relative to another along their mutual surface can be described in terms of 'Pole of rotation.' The Pole of plate rotation is completely independent of the Earth's spin axis and has no relation to the magnetic Poles. Plate velocity varies all along the sphere of the Earth. Maximum velocity occurs at the Equator of rotation and minimum velocity at the Poles of rotation.

Each of the plates are moving at different speeds. The rate of plate motion can be measured by dating of magnetic reversals or the linear magnetic anomalies found on the ocean floor. During the last 2 b.y., the Earth's magnetic field has reversed from North to South. The reversal although irregular (some as short as 20 thousand years, others as long as 10 m.y.) can be known by radiometric dating and the establishment of isochrons (magnetic time lines). Current plate motion is measured by using satellite and lasers.

A laser beam from the Earth is beamed to a satellite whose position is known very accurately. The satellite reflects the beam, which is collected at the surface of the Earth again, and the elapsed time is determined. This allows the location of the laser to be determined to within a centimetre. If the location of several such stations on different plates are repetitively determined, the relative motion between plates can be accurately measured.

Plate Movement Regarding Each other: Since each plate has a high flexural rigidity, it moves as a single unit, i.e., if one part of the plate moves, the entire plate moves. It can be warped or flexed slightly as it moves but relatively little changes occur in the middle of the plate. Almost all tectonic activity occurs along the plate boundaries and is associated with different motions between adjacent plate. Three types of plate boundaries and margins are identified which define three fundamental kinds of deformation, tectonic and geological activity—Constructive Margin or Divergent Plate Boundary, Destructive Margin or Convergent plate Boundary and Conservative Margin or Parallel Plate Boundary or Transform Fault Boundary.

Plate Boundary and Plate Margin: A plate boundary is the surface trace of the zone of motion between two plates. A plate margin is the marginal part of the plate. Two plate margins meet at a common boundary.

Constructive Margin: Constructive Margins are rift zones or spreading zones. They are zones of tension where the lithosphere splits, separates and moves" apart (diverges in two opposite directions) as hot magma comes up through cracks and solidifies leading to the formation of new crust.

Plate divergence takes place when the lithosphere under tension splits along a rift (rift zone) and the fragmented portion moves apart in two opposite directions as hot magma wells up from beneath through cracks and solidifies leading to formation of new crust, hence known as constructive margin. It is this mechanism which was responsible for the break up of Wegener's Supercontinent Pangaea into India, Australia, Antarctica, Africa and the Americas. Huge continental masses may thus break and move in different directions creating ocean basins in the process. To crack the lithosphere under the continents, it requires a mechanism to thin and apply tension to the plate of which continents are a part. The process of continental rifting and ocean basin formation involves a sequence of development stages which are—Intracontinental rifting, Interplate thinning and Ocean ridge formation.

Intracontinental Rifting: The fragmentation of a continent is initiated by an upward movement of convection currents, in the form of rising plume of molten rock from below. The effect of this activity is that the crust directly above is heated, bowed, upwarped and experiences domal uplift. Since a plate is a rigid body upwarping causes stretching of the crust and generates numerous tensional cracks and developments of rift starts. As the convection currents in the form of rising magma (plume) spreads laterally, the broken portion is pulled apart. Gradually, the broken slabs slide downward into the gap created by diverging plates creating a rift valley or graben. Intracontinental rifting was the first stage in the fragmentation of the Supercontinent. Possible present day examples are the East African Rift Valley systems (which are narrow and long), Rhine trough and Baikal (narrow and short).

The East African Rift valley running from Mozambique to Ethiopia is formed by doming and cracking of the lithosphere caused by a rising mantle plume. It represents the initial stage in the break up of a continent. The extensive volcanic activity believed to accompany continental rifting is exemplified by large volcanic mountains such as Mt. Kilimanjaro and Mt. Kenya.

Interplate Thinning: As hot magma continues to rise beneath the lithosphere, partial melting takes place. The effect of partial melting is to thin the lithosphere. As thinning proceeds, the surface sinks and the faults deepen, gradually allowing magma to penetrate into the graben, forcing the crack to widen and deepening the rift. If the rift deepens sufficiently, sea water may enter. The Red Sea today is at this intermediate stage with both oceanic type basalt and subsided block of continental type crust present on the floor of the rift. Its formation has helped Arabian Peninsula to part from mainland Africa. The Red Sea is basically an extension of the East African Rift valley but it is a stage ahead of the East African Rift valley.

There was no such thing as Red Sea a few million years ago. It was a rift valley very much like the East African Rift valley and before that, part of African continent. If the spreading process continues in East Africa, the rift valley will lengthen and deepen, eventually extending out into the ocean. At this point, the valley will become a narrow line or sea with an outlet to the ocean similar to the Red Sea today. East Africa will then eventually part from the mainland in much the same way the Arabian peninsula did just a few million years ago. The Gulf of California is another such region that is opening up.

Ocean Ridge Formation: As the crustal plate continues to thin at the rift, magma wells into the rift at an increasing rate and the landmass is gradually separated into two parts with a low lying region of oceanic basalt type crust between the two sections forming a new ocean basin. Magma continues to come through the cracks and solidifies forming ridges on either side. The Carlsberg Ridge apparently continuous with the Red Sea structure is in the third stage of ocean basin formation. The Mid Atlantic Ridge actually formed when both Africa and South America and Europe and America separated from each other by rifting of the Supercontinent Pangaea. The rift still present in the form of Mid Atlantic Ridge provides channel ways for magma to ooze up and form ridge. Similarly, East African Rift valley is the site of a narrow linear sea and still ahead an ocean basin occupied by MOR. It is not that all rift valleys develop into full fledged spreading centres. The rift zone extending from Lake superior to Kansas is an aborted rift zone. The Benue trough and the Narmada Son Damodar lineament are other examples. Why one rift valley continues to develop while others are abandoned is not yet known?

Divergence beneath oceans result in:

1. Volcanic activity,
2. Formation of new crust, formation of submarine mountains, ridges and rises whose crest is occupied by a rift valley, and
3. Occurrence of shallow foci earthquake.

As the oceanic plate diverges, a rift is created which upsets the pressure-temperature equilibrium causing the partial melting of upper mantle which is composed of peridotite.

This melted portion comes on the surface in the form of a basaltic eruption. It solidifies and is welded on either side of the plates forming new crust and pushing the plates apart. Successive eruptions of basaltic lava leads to the formation of submarine mountains or oceanic ridges and rises. If the plates are spreading fast, the oceanic ridge is very gentle and smooth, e.g., the East Pacific Rise.

When plate separation is slow, the oceanic ridges are relatively steep and rugged. An intermittent plate separation such as that associated with Atlantic, results in the magma chamber beneath becoming hollow during the period of quiescence leading to sinking of the overlying surface and in the process creating a rift valley. Rift valley formation seems to be a normal process as the movement of plates on either side creates gaps. To fill this gap, slabs slide downward.

The large downfaulted valleys generate rifts. The ocean ridges form a discontinuous mountain chain, which practically girdles the Earth. The crest of this mountain chain is occupied by rift valleys. All along the oceanic ridges, the stretching is not uniform. This differential stretching is because the plates are moving, along a Pole of rotation, with minimum velocity at Poles increasing equatorward. Thus, the oceanic ridge is offset by many transform faults. Movement along these transform faults generates earthquakes, which have a shallow focus.

Plate Boundary that is Convergent

Plates are said to converge when two plates from opposite direction come together and collide. Upon collision, the leading edge of one (the plate having higher density) is bent downward, allowing it to descend beneath the other. Upon entering the hot asthenosphere, the plunging plate is heated, melted and is completely assimilated within the material of the upper mantle. Since one of the plate is destroyed here, this boundary is known as a destructive margin.

Although, all convergent zones are very much similar, the nature of plate collisions is influenced by the type of crustal material involved. Collisions can occur between two oceanic plates, one oceanic and one continental, or two continental plates.

Occurrence of Ocean-Ocean Collision: When two plates, having oceans at their margins, collide, the resulting phenomena is known as ocean-ocean collision, e.g., the collision of the Pacific and the Eurasian plate. As a result of convergence, the oceanic plate having higher density descends into the asthenosphere. This process called subduction and this region is called the subduction zone. The zone of ocean-ocean collision is a zone of deformation, metamorphism, Earthquake, volcanic activity, island arc and trench formation.

As the two plates carrying ocean collide, a trench is formed as the plate descends. Where the Pacific plate is colliding, it has produced the Kurile, Japan, Philippines, Mariana and other trenches. The region of collision of two plates is a zone of very high pressure and low temperature. The sediments are thus metamorphosed giving rise to a very distinctive assemblage of rocks. These rocks formed under high pressure and low temperature are known as Blueschist metamorphic rocks.

Subduction zones are the sites of the most widespread and intense earthquakes. Although, the surface characteristics of the earthquakes associated with oceanic trenches and island are varied, the majority of such Earthquakes appear to be confined to a narrow dipping zone which has become known as the Benioff zone. Benioff projected earthquakes from a typical trench area on to a single vertical plane and viewed it in profile. It is seen that a broad band of epicentres at the surface, some hundreds of kilometres wide, correspond to a rather narrow inclined zone of seismicity defined by Earthquake foci. This zone meets the surface close to the line of deep ocean trench and dips away from the ocean towards the continent or, island arc at about 45°. Tensional earthquakes occur on the oceanic side of the oceanic trench where normal faulting occurs from tensional stresses generated by the initial bending of the plate. Shallow earthquakes are produced by dip slip motion resulting in thrust faulting as descending plates slide beneath the overlying plate.

This type of activity persists up to a depth of 100 km. At intermediate depths, earthquakes are caused by extension or compression depending on the specific characteristic of subduction zone. Extension and normal faulting result when a descending slab, which is denser than the surrounding mantle, sinks under its own weight. Compression results, if the mantle resists the downward motion of the descending plate. The zone of deep earthquakes shows compression within the descending slab of lithosphere indicating that the mantle material at that depth resists the movement of the descending plate.

Although, most seismic activity is restricted to the Benioff zone, more irregular activity occurs in the overlying mantle and crustal rocks. Most of the Earthquakes away from the Benioff zone are associated with the rise or formation of magma which eventually feeds the island arc or mountain chain bordering the oceanic trenches.

The subducting plate finally gets melted. Melting increases the volume, and thus, the melted material finds its way upward in the form of volcanic activity. This melted material piles up continuously on the adjacent ocean floor forming a volcanic island. This volcanic activity alongwith the buoyancy of the intrusive igneous rock, emplaced within the crust below, gradually increases the size and elevation of the volcanic islands. Since these volcanic islands are arranged in an arc shape, they are known as island arcs. Erosion of island arc starts and the sediments reach the trench where they are metamorphosed by compressional forces. All this results in the formation of complex system of volcanic rocks, folded and metamorphosed sedimentary rocks and intrusive rocks.

One of the most interesting aspects of subduction is that the back arc areas or basins (sea floor behind island arc) are not squashed by the collision of plates, rather they seem to be extending. It happens that as the oceanic plate is subducting, it causes some turbulence in the flow of asthenosphere. As a result of this, complex convective eddies are formed in the asthenosphere above the subducting plate and beneath the back arc basin. This creates regional tension, stretching, crustal thinning and block faulting. All this ensures that the back arc basin will extend and spread. Back arc spreading is taking place in the Sea of Japan, Lau basin, Sea of Okhotsk, Philippines Sea, China Sea and behind Tonga Trench.

In the western Pacific Ocean, where the Pacific plate is getting subducted, double trenches are found. This is because of abandonment of one subduction zone and formation of another. This process has occurred several times, and hence, double trench phenomenon is common.

As the oceanic plate approaches the subduction zone, some of the deep marine sediments on the oceanic plate can be crumpled and deformed. Slabs of oceanic crust along with the sediments are scraped off by the overriding continental material and are incorporated in a chaotic mass of complexly disturbed rocks called a melange (French: 'Imelange'-a jumble). In this way, formerly deep ocean rocks can be thrust up in mountain masses and can be exposed at the Earth's surface. This process is called obduction.

Within the melange, distinctive assemblage of deep-sea sediments, submarine lavas, peridotite and gabbro all together form an ophiolite suite.

The thick sequence of rocks along the continental margins is then compressed and deformed to form a mountain range like the Andes. As the oceanic plate subducts, earthquakes are generated in the Benioff zone. As the oceanic plate plunges deeper, it gets melted, the partial melting of mixture of basaltic rocks and sediments generates andesitic magma and occasionally granite. Most of the rising magma will be emplaced in the overlying continental crust when it will cool and crystallise at a depth of several hundred km forming batholiths. The remaining magma may eventually migrate to the surface where it gives rise to numerous and occasionally explosive volcanic eruptions. Thus, where Nazca plate subducts, the molten material is thrown up by Cotopaxi, Chimborazo and Misti volcanoes; where Juan de Fucca plate subducts the same material is thrown up by Mt. St. Helens volcanoes.

Occurrence of Continent-Continent-Collision: Continent-continent collision occurs when two plates carrying continental crust collide, as is the case with the Indo-Tibetan collision. Continent-continent collision is characterised by mountain building, ophiolite suite, earthquakes and remnants of past volcanic activity.

Prior to continental collision, the landmasses involved are separated by the ocean crust. As the two plates converge, the intervening sea encloses and sea floor subducts beneath one of the plates, the partial melting of the descending oceanic slab and the sediments carried with it, generate a volcanic arc. If the subduction developed at a substantial distance into the ocean, an island arc would form. The already sediment laden continental margin would receive further sediments by erosion of the newly formed volcanic island. As the ocean basin continues to shrink by consumption of the sea floor, these sediments will get squeezed, deformed and folded, predwing a mountain range composed of deformed sedimentary rocks and fragments of volcanic arc.

Continued convergence results in the scraping off a layer of oceanic material (obducted), which gets plastered on to the other plate in form of an ophiolite suite. Ophiolites mark the zone where two plates have got stitched, therefore, it is also known as suture (Latin: *'sutura* '-stitch) zone.

Finally, the continental margins of the other plate come close enough to collide. In this case, the low density of continental material does not permit its subduction. The oceanic material breaks from the continental block and gets completely assimilated in the mantle. Volcanic activity now ceases. As the buoyancy of the continental lithosphere does not permit it to go very far in the mantle, continued convergence forces the plate partially under the plate creating an unusually thick layer (in fact two layers) of continental crust that supports high rising topography. Since the continental crust can never be completely destroyed, whenever the continental part of the plate arrives at the subduction zone, it may lead to some changes in the pattern of interplate motion between the two plates. Alternatively, this may also involve adjustments in the motion of the other boundaries of the two plates and the formation of a single plate from the two, which previously formed.

The Alpine Himalayan chain was formed in this manner. The advancing Indian plate enclosed the Tethys Sea and crumpled its sediment into a huge mountain mass. A portion of Indian plate was underthrust beneath the Eurasian landmass creating a double layer of low density supporting the high elevation of Tibetan plateau. The Mediterranean Sea, is in fact, a remnant of Tethys Sea. Here, the oceanic plate has not been fully consumed and the margins are still in the process of consumption. Where the Arabian plate subducts with the Asiatic plate, it sends quakes that plague Iran and Turkey, where the African and European plate meet, the African plate subducts and it is the melted material in this subduction that throws volcanic material from Etna, Vesuvius and Stromboli.

Thus, destructive margins are the site where plates get destroyed. This is necessary because new crust is being constantly generated at ocean ridges and if the generation of new crust where not balanced by destruction, the Earth's surface area would have increased. This, however, is not the case. The situation in terms of Africa and Antarctica are entirely different. Both of these continents are surrounded by ridges and have no associated subduction zones to accommodate the new lithosphere generated at the ridges.

Although, the African plate is moving northward towards the convergent boundary in the Mediterranean area, it is not able to accommodate the east-west spreading from the Atlantic and Indian ridges. Same is the case with Antarctica plate. Basically, the African and Antarctica plates are being enlarged as new lithosphere is being created at their margins. Since there are no subduction zones, the ridges which surround this plate must be moving outward in relation to Africa and Antarctica. The plates are, thus constantly enlarged, so the ridges themselves must move. Continued movement of ridges will ultimately lead them towards subduction zone where both the ridge and subduction zone is finally destroyed and a single plate is formed. Since the ridges are moving, the African plate must be growing. Indeed, the plates in the Pacific appear to be decreasing in area to accommodate the growth of African plate and other growing plates. If the current rate of shrinkage of Pacific basin continues, the American continents may collide with Asia in 200 m.y.

Conservative Margin: At conservative margin, the plates slide past each other without the formation of new crust. This boundary is, thus, also called parallel or transform fault boundary. Transform faults roughly move parallel to the direction of plate movement. Transform faults are so called because the kind of motion between plates is changed or transformed at the ends of the active part of the fault, for example, the diverging motion between plates at an oceanic ridge can be transformed along the fault to the converging motion between plates at subduction zone. Transform faults connect convergent and divergent plate boundaries in various combinations.

Transform faults are dominantly found on the oceanic crust where they offset the oceanic ridges throughout their length of 64,000 km. The non-oceanic transform faults are the San Andreas fault, California; the Alpine fault, in New Zealand; the Philippines fault; and the Atacama fault, Chile. San Andreas fault is a transform fault that slashes through the continental crust. It is a ridge to ridge transform fault that connects the East Pacific Rise and Juan de Fucca Ridge. Juan de Fucca Ridge moves in a southeasterly direction and eventually subducts under the west coast of the United states. The southern end of this relatively small plate is bounded by the Mendocino escarpment. The fault facilitates the movement of crustal material generated at oceanic ridges to its destruction beneath North American plate. It is along this fault that the Pacific plate is moving in north-westwardly direction.

Plates will converge, diverge or move parallel to each other but the only way they can end is in the form triple junction. Triple junctions are points where three plates meet. Such junctions are the necessary consequence of rigid plates on a sphere since this is the only way a plate boundary can end. There are 16 possible combinations of ridge, trench and transform fault, triple junction of which only six are common. These triple junctions can be either stable or unstable depending on whether they preserve their geometry as they move.

Movement of Plate

There are various forces that act upon the plates to make them move. Three classes of mechanisms are possible:

Convection Currents: Plates can be driven by convection currents. Three forms of convection are possible.

Convection Involving the Whole Mantle: There is a large amount of radiogenic heat (100 joules/year) concentrated in the mantle. This leads to the formation of convection currents. As the currents ascend from below, they diverge and spread laterally. Convection would cause the lithosphere to split and as the plates may move laterally, the currents carry the overlying slab of lithosphere with them. This marks the site of ocean ridges. The currents encounter a similar current from the opposite direction and descend into the deeper parts of the mantle and drag the lithosphere down into the mantle at trenches. The high heat flow at oceanic ridges is an evidence to this process. There are, however, two major problems—(i) that the mantle is layered and the sort of convection would preclude such layering, (ii) below 700 km mantle is too resistant to allow penetration of subduction plates.

Convection Involving only Asthenosphere: In this case, the size of convection cells is smaller.

Mantle Plume: Another variety of mantle convection involves jet-like plumes of low density material from the core mantle boundary. As the plume reaches the lithosphere, it spreads out laterally doming surficial zones of the Earth and moving them along in the directions of radial flow. The centre of the Afar triangle in Ethiopia is one site of plume that flowed upward and outward carrying the Arabian, African, and Somali plates and moved away from the centre of the triangle. Geologists think that other such triple rift system in the past had a similar origin. Mantle plumes are thought to be responsible for the breaking of the Pangaea.

Slab Pull: Slab pull is thought to operate at subduction zones where the subducting oceanic plate becomes colder and denser than the surrounding mantle (here gabbro metamorphoses into eclogite) and pulls the rest of the slab along. It is believed that the absolute velocity of a plate chiefly depends on the proportion of the margin that is subducting. This subduction is aided by gravity pull, for example, plates like Pacific and Cocos, which have about 40 per cent of their margins represented by subduction zones, have high plate velocities (5 cm/yr.). Plates like the North American, which have smaller proportion of subducting margins, move more slowly (1 to 3 cm/yr.). Hence, it can be said that the major driving force is slab pull.

Ridge Push: Ridge push results from the fact that:

Spreading centres stand high on the ocean floor. This result in gravitational sliding of the lithospheric slab away from the oceanic ridge raised by rising material in the asthenosphere. It has been estimated that a surface slope of only 1: 3000 would produce a movement of the lithosphere of 4 cm/yr. over sathenosphere.

Magma rising along the axis of the zone of spreading form wedges of new lithosphere on either side of trailing edge of the plate, thus the plates are pushed apart.

Concept of Hot Spots and Mantle Plumes: About 100 small regions of isolated volcanic activity scattered throughout the world are known to geologists as hot spots. These hot spots are surface manifestations of mantle plumes. Mantle plumes are isolated long slender columns of hot rock that originate deep inside the Earth's mantle. They are thought to have diameters of 100 to 240 km and rise slowly at rates of about 2 metres per year. These hot spots, which channel heat to the surface from deep within the mantle, are fixed spots and erupt periodically. Hot spot plumes of magma are not uniform, they differ in chemistry, suggesting that they come from different depths. They last for about 100 m.y.

At least 122 hot spots have been active in the past 10 m.y. Several of them lie on mid-oceanic ridges or close to them. The most prominent are Iceland, Azores and Triston da Cunha. The amount of material ejected by these hot spots exceeds the norm for mid-oceanic ridges, that is, they have built up into islands while the rest of the ridge crest remains submerged. The lavas, extruded by these hot spots, are typically alkali rich basalts, rare at plate margins but typical of hot spots.

Of the 122 hot spots, 69 are on continents and 63 in ocean basins. Among the oceanic hot spots, there is tendency to congregate on mid-ocean ridges. 15 lie on the crests of ridges and nine others are near the crests. African plate, however, has the greatest concentration which has 35 per cent of hot spots in 12 per cent of world's surface area. The large scale topography of African continent is characterised by basin and swell topography, the unusual topography and the number of hot spots are entirely related and can be explained by the hypothesis that Africa has come to rest over a hot spot. The most compelling evidence that Africa is stationary is that at some hot spots lavas of several ages are superimposed.

It the continents were moving then these lavas would be spreading out in a chronological sequence as has happened in Pacific with Emperor Sea Mount chain, Austral-Marshall-Gilbert chain, Line Tuamotu chain. As the oceanic crust is moving over a hot spot, successive eruptions can produce a linear series of peaks or seamounts on the moving crustal plate. In such a series, the youngest peak is above the hot plume and the seamounts increase in age as the distance from the hot spot increases.

Thus, Hawaiian island, is an active volcano located at the end of Emperor Seamount chain. Maui, an island about 50 km is hardly extinct. The change in the direction of chain of islands indicates change in the direction of plate movement. The Emperor Seamount chain began to form more than 40 m.y. ago when the Pacific plate was moving northward. About 25 m.y. ago, the plate moved northwest ward and started to form Midway-Hawaiian chain. Similar trends are shown by Austral-Marshall Gilbert and Line-Tuamotu chain. In the Atlantic, the Walvis ridge and the Rio Grande ridge were formed similarly over Tristan da Cunha hot spot when African plate migrated eastward and South American plate westward over this spot. Since the lithosphere was very thin here the hot spots, it produced a continuous series of seamounts very close together forming a transverse ridge Walvis to the east between Africa and Mid-Atlantic-Ridge and Rio Grande to the between South America and Mid-Atlantic Ridge.

Problems and Theories of Plate Tectonics

1. While the lateral extent of modern plate can be easily defined by earthquake activity, its thickness is not physically defined. This lack of definition makes it difficult to understand how plates move.
2. Did plate tectonics occur in the past? Even if it occurred, was its manifestation different from that in present day? These questions are not easily answered. The main problem encountered in interpreting the past evolution of plate margins stems from the fact that the crust disappears at the subduction zones thereby destroying half of the history of evolutionary sequences. Ophiolites can assist the recognition of a vanished ocean along the sutured collision zone of two continental masses. These, however, do net reveal the previous width of the ocean. Moreover, ophiolites can be remnants of flood basalts erupted on the continental blocks with no associated ocean at all. Where there was any ocean or not, the record of this sort will be entirely destroyed, if the ocean ridges subduct.
3. Lack of exact symmetry of aseismic ridges even when spreading of the ocean floor was symmetric.

4. Subduction of oceanic ridges is puzzling. If the subduction is due to drag, then the descending slab would cease to be coupled when the ridge reached the subduction zone, and hence, the far side of the ridge would not subduct.
5. It is difficult to imagine how a ridge over the rising limb of a mantle convection cell would be carried down on the descending limb of the same cell.

Significance of Plate Tectonics

Despite these problems, there is no denying the fact that plate tectonics has provided a single unifying theory for the understanding of a wide variety of Earth phenomena, which were previously neither understood in themselves nor in any wider context. Plate tectonics has been responsible for explanation of almost all features on the Earth's surface, continental as well as oceanic.

Plate Tectonics has brought a revolution in Earth Sciences. What Copernicus theory did to astronomy, the theory of plate tectonics has done to Earth Sciences. Although, the continental drift theory did provide the initial insight into many of the processes operating on the surface. Plate Tectonics theory has been an enormous improvement over the continental motion floating lithosphere over continental drift gravitational and drift theory. Continental drift said that only the continents are in motion while according to Plate Tectonics not only the continents but oceans are in motion too. According to Plate Tectonics theory, it is not SIAL that is in over the SIMA but the asthenosphere. The theory talked about the tidal forces as the chief mechanism for the movement of the continents which were largely inadequate and preposterous while Plate Tectonics theory gave five different mechanisms for the movement of the continents.

The explanation given by Plate Tectonics theory on every aspect, which the continental drift hypothesis touched, has been an improvement. The basic merit of the plate tectonics concept lay in the fact that it has been able to explain every feature and every process found on the crustal surface so convincingly that it is no more considered to be a theory—it is a fact.

Difference between Plate Tectonics and Continental Drift

	Continental Drift	Plate Tectonics
1.	Continental drift is a theory proposed by Alfred Wegener. theories, concepts and processes in the 1970's.	Plate Tectonics theory is the culmination of many SIMA and NIFE.
2.	Continental drift theory assumes that the Earth can be divided into SIAL,	Plate Tectonics process divides the Earth into Lithosphere, Asthenosphere and Mesosphere.
3.	According to Continental drift theory, continents are made of SIAL (Silica and Aluminium) and are floating over SIMA.	According to theory of Plate Tectonics, plates are hard rigid segments of the lithosphere floating on the asthenosphere.
4.	According to Continental drift theory it is only the continents that are in motion.	According to theory of Plate Tectonics, it is not only the continents but the oceans as well are in motion.
5.	Continental drift envisages that the continents broke up from the ancient Supercontinent Pangaea but it does not answer the question whether the broken fragments will reunite.	Plate tectonics envisages the fragmentation of a single Supercontinent and their reassembly to form a Supercontinent again.
6.	After fragmentation the broken portion of Pangaea drifted northward and westward.	After fragmentation the broken portion of plates move in all direction.
7.	Continental Drift theory does not stipulates stages for the break up of the continents	According to theory of Plate Tectonics, the break up of the continents is achieved in various stages. Intra Continental Rifting, Interplate Thinning and Ocean Basin Formation with ocean ridges.
8.	Continental drift stipulates for simple break up and its movement without any mention of types of margins.	The movement of plates leads to three types of margins-divergence, convergence and parallel motion, all different in their mechanism.
9.	The mechanism of continental drift is tidal forces and gravitational pull of the Earth.	The mechanism of plate movement include convection currents involving either the whole of mantle or asthenosphere, mantle plume, slab pull or ridge push.

Plate Tectonics and Rock Cycle

Rock cycle can be best described by plate tectonics. Weathered material from elevated landmasses is transported to the continental margins where it is deposited in layers that collectively are thousands of metres thick. Once lithified, these sediments create a thick wedge of sedimentary rocks flanking the continents. In due course of time, sedimentation along the continental margin may be interrupted, if the region becomes a convergent plate boundary. When this happens, the adjacent oceanic lithosphere begins to subduct beneath the continent into the asthenosphere. Along such active continental margins, the converging plate deforms the margins of the sedimentary rocks and transforms them into linear belts of metamorphic rocks. As the sedimentary rocks descend further, some of the overlying sediments that were not crumpled into mountains, are carried downward into the hot asthenosphere, where they too undergo metamorphism. Eventually, some of this metamorphic material will be transported to depths where the temperature and pressures are sufficiently great to initiate melting. This newly formed magma will then migrate upward and erupt on the surface. The magma, in its course of movement upward, solidifies within the rock or above the surface to form igneous rocks. These igneous rocks are immediately attacked by the process of weathering and the process is renewed.

Continental Drift: Brief Account

There are a number of features found on the Earth's crust resulting from two forms of energy—internal energy and external energy. The stresses set by meteorites and the presence of radioactive elements all provide internal energy. Internal energy is responsible for creating endogenic movements. The endogenic movements are, in turn, responsible for gigantic movements on the Earth's crust. Plate tectonics is the manifestation of such a movement. Plate tectonics imply large scale horizontal and vertical motion where the exogenic processes also play an important part in maintaining equilibrium.

Concept of Continental Drift

The suggestion that there might have been lateral displacements of the continental masses on a gigantic scale is generally ascribed to Alfred Wegener (1910). But the germ of the idea can be traced back to 1620 when Francis Bacon observed the parallelism of the opposing coasts of the Atlantic. Frenchman P. Placet made similar types of observations in 1668. In 1858, Antonio Snider united the continents. Snider's reconstruction of Carboniferous Geography was to explain why the fossil plants preserved in the coal measures of Europe are identical with those of North American coal measures. Snider's observations, however, were soon forgotten. F. B. Taylor, while accounting for the distribution of mountain ranges, visualised the Laurasia stretching and splitting while moving from the Poles to the Equator. Similarly, the Gondwanaland also ruptured and splitted. Because of lack of a suitable mechanism, this theory was not accepted. These works may have prompted Wegener to postulate his famous theory of continental drift.

Wegener's theory basically grew out of the need to explain the major variations in the Earth's climate. He sought the solution of this problem either in the movement of landmass or the change in climate but having gained no evidence of climatic changes. Wegener proposed the movement of continents or drift of continents as the reason behind climatic change. At the same time, Wegener was also struck by the congruence of Atlantic coastlines and the intellectual problems raised by the 'land bridge' explanations of links between continents seemed inseparable. Such land bridges were imagined island chains or continents which for long had foundered and it facilitated the movement of animals and plants in the past and could thus explain the occurrence of similar creatures in widely separated continents.

Wegener began by assuming that there was a super continent called Pangaea, surrounded by an ocean called Panthalasa. The northern part of the Pangaea was called Angaraland while the southern part was called the Gondwanaland. The northern and southern part was separated by a geosyncline, a long narrow shallow sea called Tethys, in which sediments were deposited. The continental crust SIAL was moving over the SIMA. The deep sea floor was taken as the upper surface of SIMA composed largely of basalt. This Supercontinent splitted during the Carboniferous period and started drifting. Drifting took place in two directions – northward under the gravitational and rotational forces and westward under tidal forces. The larger and sturdier continents actually broke through the oceanic crust much like the icebreakers cut through ice.

Wegener cited various evidences to show that the continents could have been close together in the past. These were—

1. The Juxtafix of continents.
2. Structural and stratigraphic evidences.
3. Fossil evidences.
4. Palaeoclimatic evidences.

Juxtafix of Continents: There is a remarkable similarity between the coastlines on opposite sides of the south Atlantic and this could have happened only when the continents were once joined together and since then they fragmented. The original jigsaw fit of the continents was rather crude but it conveyed these ideas.

Structural and Stratigraphic Evidences: If the continents were once joined together, the rocks found in a particular region on one continent should closely match in age and type with those found in adjacent positions on the matching continent. Indeed, in northwestern Africa and eastern Brazil both 550 m.y. old rocks lie adjacent to rocks dated more than 2 b.y. in such a manner that the line separating them is continuous when the continents are brought together. Mountain belts terminate at one coastline and reappear again on a landmass across the ocean. These includes the Brazilides and Gondwanides which terminate on South American coast and reappear again in South Africa, the Appalachian trends—northeastwards through the eastern United states and disappears off Newfoundland coast. Mountains of similar type and similar age (Caledonian and Hercynian) are found in Greenland and northern Europe.

Fossil Evidences: If an organism is found on two different landmasses separated by a vast ocean, there are only two possibilities—either it must have swam across (in which case it should be having wide geographical distribution) or else it is found on continents which have since drifted and thus, have a restricted geographical area. *Mesosaurus,* an aquatic reptile, fossil remains of which are found only in eastern South America and South Africa. If *Mesosaurus* had been able to swim across vast ocean, it should have been widely distributed. But, as this is not the case, South America and Africa must have been joined. The fossil fern, *Glossopteris,* is widely distributed in Africa, Australia, India and South America (later discovered in Antarctica also). These fossil ferns, which grew only in subpolar climate, are today found in warm climatic regions separated by wide oceans. Thus, it can be inferred that they were associated with a Supercontinent where fairly cool climatic conditions existed. The existence of coal deposits in presently cool climatic conditions of the northeastern USA and parts of Britain is only because of continental drift. Coal deposits could only have formed when the continents occupied Equatorial location. Later, they must have drifted and occupied a different geographical location.

Palaeoclimatic Evidence: This is the most convincing line of evidence for Wegener's work. Layers of glacial deposits (tillites) and striations and grooved marks are found in tropical regions of South Africa, South America, Australia and India (Talchir). This can be explained in two ways:

1. The Earth went through a period that was cold enough to generate extensive continental glaciation. This, however, is not the case. This is because at that time, large swamps existed in the Northern Hemisphere. These swamps with their thick vegetation became the major coalfields of the eastern United States, Europe and Siberia.
2. The landmasses were joined to form a Supercontinent and South Africa was centred over the "South Pole. This is further proven by the direction of the ice flow, which shows a radiating pattern in all directions away from southwest Africa. The scarcity of glacial deposits in South West Africa coupled with abundant evidence of erosion can be interpreted as meaning that southwest Africa was covered by an actively eroding ice sheet which dumped its load farther where glacial tills strewn all over the surface is found.

These ideas met with varied responses at the time, but most were hostile and many ridiculed his concept. Every point made by Wegener, including the Atlantic 'fit', was questioned. Particularly, the geophysical mechanism envisaged for the movements was considered inadequate. The forces tentatively suggested by Wegener to account for the movement (a rather vague tidal force combined with movement towards the Equator due to rotations) were insufficient and that in any case it was proven that the rocks of the continental crust/ocean crust division were rigid and would not permit the drifting of the continents over ocean floor. Further, the ocean floor was not so weak that it permitted passage of the continents without themselves being appreciably deformed.

Support for Wegener came from the South African geologist, DuToit and the British geologist, Holmes DuToit amassed a large volume of evidences to show links between continents in the past. Holmes was involved in applying his studies of radioactive elements to internal energy sources and suggested a model which directed attention to a zone somewhat lower in the Earth than the continental crust/ crust boundary along which movements might take place. Most scientists, however, rejected the idea of continental drift—largely on geophysical grounds—until geophysical evidences from palaeomagnetism and ocean floor studies led to the synthesis of ideas in Plate Tectonics.

Process of Palaeomagnetism

The Earth has a magnetic field, which is similar to that produced by a bar magnet, at the centre of the Earth aligned parallel to Earth's spin axis. Certainly, there is nothing like a bar magnet in the Earth's core but it is a fact that the field behaves like a bar magnet with the lines of force emerging at the Earth's surface at specific angles for a particular latitude.

Configuration of Bar Magnet

It is not known for sure exactly why the Earth itself behaves like a magnet. The configuration of the Earth's magnetic field is like that of a strong bar magnet. It has an outer rocky mantle almost 3000 km deep. Below that. is the liquid part of the core about 2000 km thick, which surrounds the solid centre. Most scientists think that the motion of charges in the liquid part of Earth's core creates the magnetic field. Because of its great size, the speed of moving charges looping around within the Earth needs only to be somewhat less than a thousandth of metre per second. Some scientists think that these currents are caused by the

Earth's rotation. This idea is supported by many facts. Jupiter, which has a 10 hour day, has a much stronger magnetic field than the Earth. The moon, which rotates once in 28 days, has a very weak magnetic field. Venus, which rotates slowly still, has no measurable magnetic field according to space probe measurements.

Another likely source of Earth's magnetism may be heat, rising from the Earth's central core. This heat may be the cause of generation of convection currents of molten material in the liquid part of the Earth's core.

The motion of ions and electrons in this molten material would produce an electric current through the Earth's core. The rotation of the Earth's core like a disc produces the Earth's magnetic field in the same manner as self-exciting dynamo. In geologic history, an unsteady flow of heat might have caused a subsiding and momentary ceasing of convection currents, resulting in a collapse of Earth's magnetic field, Regeneration of heat would have again caused convection currents but not necessarily in the same direction. If the convection currents reversed, the Earth's magnetic field would also reverse.

When sediments containing ferromagnesian minerals are deposited, they generally become weakly magnetised. Similarly, when igneous magmas crystallise, particularly basaltic (basic sills and dykes), every crystal of ferromagnetic mineral, particularly oxides of iron and titanium such as magnetite and illmenite, acquire a stable magnetism which becomes frozen, as it cools below Curie point (the temperature, above which atomic magnets are free to rotate and below which they are 'frozen' in orientation and is normally 600°C). The acquired magnetism, now called fossil magnetism, has the same direction (declination D) and the same dip (I) as the local geomagnetic field at the time of consolidation. The Earth's magnetic field generally points northward. It intersects the surface at various angles depending on the latitude. It parallels the surface at the Equator, points outward from the surface in the Southern Hemisphere and points to the surface in the Northern Hemisphere.

The angle it makes with the horizontal is called the magnetic dip. This is zero at the Equator but increases with latitude becoming vertical near the Poles. By measuring the dip of fossil magnetism in ancient lava flow, the latitude can be determined. The basic premise on which palaeomagnetic investigations are based is that the geomagnetic field averaged over an appropriate time has always been very nearly that of geocentric axial dipole, i.e., the geomagnetic Pole has, on an average, coincided with the geographical Pole. If this coincidence were exact, then for rocks of a particular age from a given locality, the averaged value of I (palaeomagnetic inclination or dip) would give the latitude X of the locality at that time by the equation, tan 1 = 2 tan X.

If both distance and direction of the geographical Poles are known, the position of Poles on the globe at that time can be determined. This method is capable of giving palaeolatitudes to within ten degree or better. However, it does not tell anything about palaeolongitudes. When palaeomagnetic evidences for a given region was checked over long periods of time, a gradual change in latitude and direction was detected. In other words, it means that the position of magnetic Poles has moved steadily over time—not simply around geographical Poles but over thousands of kilometres. This movement can be plotted on the map to give polar wandering curve. These curves are merely lines on a map connecting Pole positions relative to a specific continent for various times during the geologic past. As the position of magnetic Pole is linked to the geographical Pole, what this really means is that the continents have changed positions on the face of the Earth, relative to the axis of rotation.

If the position of the Poles for rocks 200 m.y., 300 m.y. and 350 m.y. old is determined, it is found that this does not coincide with the present day position of the Poles. Two inferences can be drawn from this:

1. That continents have remained *'in situ'* and the Poles must have changed their position. However, it is impossible that there were many Poles migrating systematically and eventually merging. It is only logical that there has been one and only one magnetic Pole, which has remained fixed.
2. That polar wandering took place because of continental drift. Assuming the continents to have remained fixed there would be a single polar wandering curve. If all the continents showed the same Apparent Polar Wandering (APW) path,

it would be clear that they did not move relative to each other and that the Poles actually moved. The fact, that every continent has its own polar wandering curve and these are widely different, means that continents have drifted. It is very much proved that if relative position of the continents changed then the Poles, determined by the contemporary rocks of two different continents, must not be the same. This fact can be explained in another way. Till two continents are joined together or are not moving (drifting) in relation to one another, the polar wandering curve will be the same. For example, the polar wandering curve of South America and Africa in their present day positions appears to be different, yet if these continents are put together the polar wandering curve joins suggesting that for some time at least the pieces moved as one.

Theory of Plate Tectonics and Rock Cycle

According to the theory of Plate Tectonics, crustal evolution takes place as a continental nuclei emerges from the upper mantle, undergoes gradual accretion and grows slowly in geological time.

The primitive crust was of an oceanic type and the continents were small or possibly non-existent. Then, the continents started growing slowly through the chemical differentiation of the mantle.

The Earth has been cooling since it first differentiated from others planets The temperature, in the Earth's early stages of history, was very high. Largely this was a result of:

1. The kinetic energy released by the impact of meteorites and asteroids.
2. The release of gravitational energy during the separation of the core (sinking of metal phase).
3. The radioactive decay of isotopes.

The larger amount of heat availability meant that the hot mantle was less viscous, flowed more rapidly and convected more vigorously than today's mantle. There were numerous hot spots (columns of hot fluid magma) in the mantle. This heat was largely dissipated by way of formation and cooling of a basaltic layer. Thus, the primitive hot Earth may have been overwhelmed by volcanism. Eruptions may have occurred entirely beneath a global ocean, which condensed from vast clouds of gas released first during accretion and later during volcanism.

Thus, there was vigorous small scale mantle convection. Partial melting above a rising convection current in the mantle formed magma of basaltic composition, which accumulated upon the floor of the ocean and eventually rose above the sea level. This crust experienced periodic melting and was recycled back into the mantle. During this process, lighter components were separated out and distributed near the Earth's surface. Where convecting currents met and moved downwards or where loading occurred, this primitive crust was hot and light enough not to sink, and instead, it compressed, metamorphosed to greenstone and became thickened. A downward bulge was created, partially melted and this differentiated magma became granitic. Because of its low density, it injected into the rocks above (greenstones, marked by large granitic intrusion). These types of rocks are preserved in the oldest shields as deformed granite-greenstone belts. In this way, a primitive crust differentiated into silicic and granitic crust and its recycling was prevented.

The solidified lava, which was unable to sink, existed like a volcanic island arc or a small continental block. These island arcs were subjected to ordinary processes of erosion and sediments were deposited on the submerged flanks of continental nuclei. The products of this chemical weathering were altered by rising hot gases and silica rich solutions from below. Recycling of this material further produced rocks of low density which were granitic in character.

At this stage in the Earth's evolution, tectonic activity started. The granitic continental nuclei, which was formed earlier, was unable to sink. Vigorous convection currents reacted in the collision of non-granitic rocks and this helped in the subduction process which began at the base of the continents. The subduction and finally melting, which was further facilitated by the presence of oceanic water, produced silica-rich (andesitic) magma due to magmatic differentiation. These low density granitic intrusions further stabilised the continental nuclei. As a result of continued collision, the newly added volcanic material and sediment were deformed into orogenic belts. At the mountain roots, metamorphism and granitic intrusion occurred. Finally, erosion and isostatic adjustment produced a new segment of the stable crust in which igneous and metamorphic rocks, which were formed deep in mountain roots, were exposed at the surface. These rocks were subjected to weathering. The sediments derived from this process were deposited at the margin and once again the entire process was repeated. Continents, thus, grew by marginal accretion.

The following 'evidences' support the occurrence of these sequence of events—

1. The location of active island arcs.
2. The accretion of mountain ranges along the margins of continental nuclei.

However, unanswered— two questions are left

1. First, in the conversion of weathered products of basalt into granite, what happened to the excess iron and magnesium, and
2. Secondly, potassium and sodium, which constitute granite, are rare in the sediments that have formed from basalt.

The continental crust, once formed, may be subjected to intracontinental rifting, if it lies above the diverging convection currents. Continued divergence leads to the separation of continents and the formation of a new crust. Of these two crusts, the oceanic crust will be basaltic and younger.

If this model is accepted, it provides a mechanism for the formation of both continental and oceanic crusts in the Earth's early history.

The basic nature of the crust (composition and structure) and how a granitic crust evolved from basaltic crust, can be explained in terms of plate tectonics. The differentiation takes place when the basaltic crust is subducted leading to generation of a granitic crust, rich in silica and acidic matter. Similarly, the growth of crust by marginal accretion can also be explained in terms of plate tectonics. Marginal accretion takes place when plate collision leads to formation of mountains on the margins of the already formed crustal nuclei.

Earth Movements and Plate Tectonics: The theory of Plate tectonics explains why different types of Earth movements take place. Epeirogenic movements take place when a mantle plume causes the crust to bulge upwards in a dome shaped manner. Thus, the Deccan, the Colorado and the Rhine Massif are getting uplifted at a very slow pace. Continuance of this process finally leads to formation of tensional cracks on the surface, such as the Great African Rift valley. The theory also forecasts the future distribution of oceans and continents. The African Rift valley may become the site of an open ocean, if the intra continental rifting process proceeds.

Mountain Building and Plate Tectonics: The mountain belts are related to moving plates, their convergence resulting in compressional force, which finally leads to folded mountains. Thus, the Rockies, the Andes, the Alpine Himalayan belt, which are active fold mountains, were formed in Tertiary period when geoclinal rocks were deformed into fold mountains. Other mountains, such as Appalachians in the USA and Urals in Russia are zones where plate tectonics was active earlier and these mountains represent the zone of welding between two different plates.

Volcanism and Plate Tectonics: The worldwide distribution of Earthquakes and volcanoes and their occurrence at specific places is related to different types of plate boundaries. Thus, volcanic activity will take place when two plates converge, as with the Pacific and the Eurasian plate, and the Nazca and the South American plate. The melted portion of the Pacific and Nazca is thrown by the volcanoes of Mayon, Luzon in Philippines, Misti, Peru, Chimborazo and Cotopaxi in Andes. Where African plate subducts, it sends fire in Iran, Caucasus and Italy. Divergence leads to quiet type of basaltic eruption.

Plate Tectonics and Ocean Floor Topography: The theory of Plate tectonics explains many ocean floor topographic features as well.

Continental shelves can form because of tectonic processes. The continental crust stretched and thinned before tearing open, as spreading developed. This process is continuing in East Africa even today. Eventually, the continent is torn and a sea opens up like Red Sea of today. As the broken portions of the continents spread apart (trailing edge), they subside because of increased distance from the region of thermal rifting and because of isostatic readjustment, i.e., in buoyant equilibrium. These thinner portions of crust would have lower elevations than the unstretched thicker regions. The added weight of accumulated sediments increased their subsidence. These became the continental shelves and coastal plains of the trailing edges of today's continents.

Theory of Continental Slope

There are three ways in which continental slopes may originate:

1. By continental rifting.
2. By the building of a terrace seaward by waves and currents with debris brought into the sea by streams or obtained from the wave cut portion of the continental shelf.
3. By downwarping of ancient landmasses which were perhaps peneplained before being downtilted.

The most outstanding feature found on the continental slopes are submarine canyons. A submarine canyon is steep-sided and V-shaped valley with tributaries similar to those of river cut canyons on land. These are formed due to continental rifting. As the continental rifting starts, a rift valley like that of East Africa forms. The sides of the rift valleys are attacked by erosion and many canyons are cut into them. As the continents separate and move in opposite directions, the margins of the continents become submerged and the canyons are drowned. These canyons then become the channel for turbidity flows and are modified later on. Some canyons, such as those on the slopes off the Hudson and Congo rivers which are on the upper part of continental shelf, appear to be extensions of valleys on land and were apparently cut into the shelf during periods of low sea level when the glaciers advanced and the rivers flowed across the continental shelves. Then, the glaciers melted, sea level rose and the valleys were drowned. Most of the canyons whose lower parts are cut deep, however, are caused by turbidity currents. Turbidity currents are powerful currents moving at a very fast speed down the continental slope. Such currents form when great landslide of mud and sand come down the slopes. The landslides may be started by

Earthquakes or simply as a result of gravity. The landslides sweep water ahead at high speeds and mud and sand are mixed with the water. With its high speed and cutting tools, the current is a powerful agent of erosion, and thus, erodes the slope and excavates deep submarine canyons. As the flow reaches the bottom, it slows down and the sediments settle in fan-shaped manner.

Abyssal Floor Formation

Abyssal floors are formed by continental rifting and new ocean crust formation. As a new lithosphere is created at the ridge, it gradually moves away from the ridge and gets modified in many ways. As the new crust cools, it contracts and the ocean basin deepens. The superficial features of the landforms created at the ridge and the various linear hills formed due to volcanic activity, magmatic intrusion and block faulting (which become the foundations of abyssal hills) get modified and concealed beneath pelagic sediments. Finally, the abyssal hills become completely buried in sediments, thus forming the flat abyssal plains. The distribution of abyssal plains and hills substantiates this explanation. Abyssal plains occur in those places, where the sea floor does not inhibit turbidity currents from spreading sediment from the continents over the ocean floor. In the Atlantic Ocean, abyssal plains occur near the margins of the continents of North America, South America, Africa and Europe. But, there are only few abyssal plains in the Pacific because the trenches in the Pacific trap the sediments, and hence, there are only abyssal hills. The largest abyssal plains in the Pacific are found in the North-East off the Canadian and the Alaskan coast where the continental mass is not bordered by any trench. Similarly, abyssal plains occur only along the margins of Africa and India and are absent in the eastern part of the ocean where the Java trench along the continental margins acts as a sediment trench.

Theory of Oceanic Ridges and Rises

There are two general types of ridges - active spreading centre ridge and rise and inactive aseismic ridge.

The active spreading ridges are about 1000 km wide and 1000-2000 m high. If the slopes of these swells are steep, they are referred to as ridges, e.g., the Mid-Atlantic Ridge, the Carlsberg

Ridge, the Lomonosov Ridge and the Pacific Antarctica Ridge. If the slopes are more gentle, they are called rises, e.g., the East Pacific Rise.

Ocean ridges and rises are formed in zones of tension where plates diverge. As the plates break and a disequilibrium in pressure-temperature conditions is created, the lowering of the pressure in the asthenosphere causes partial melting of peridotite which results in the formation of basaltic magma. The basaltic magma ascends upwards through the crack or fissure and solidifies forming new crust. This newly solidified material is magnetised in alternative directions as the Poles reverse forming linear chain of magnetic anomalies. The constant piling up of volcanic material leads to the formation of a ridge.

The general character of the oceanic ridge seems to be a function of the rate of plate separation. Where the rate of spreading is relatively slow (<5 cm yr.$^{-1}$) the ridge is higher and more rugged and mountainous than it is where rates of spreading are more rapid. For example, the Mid-Atlantic ridge is more rugged and higher than the East Pacific rise, where the rate of spreading is very fast (>10 cm yr^{-1}).

The mountains of ocean ridges are not like mountains of the continents. While the ridge is composed entirely of basalt and is not deformed by folding, the mountains of the continent are largely made of folded and metamorphosed sedimentary rocks.

Lengthwise, along some portions of ridge systems crest, is a rift valley, about 15-50 km wide and 500-1500 m deep. These rift valleys form due to the subsidence and downfaulting of the central section as two plates diverge. In the Atlantic, where volcanic eruptions are intermittent, rift valleys form during times of quiescence when the magma chamber has been hollowed, thus allowing the upper crust to fall and subside. The oceanic ridge is traversed by a series of fractures with steep sides running perpendicular to ridges and rises displacing sections of ocean ridges. These transform faults form due to differential stretching of plates on a sphere. One of these fracture zones is the Romanche fracture zone which forms an important link for the flow of deep ocean water from western to eastern Atlantic basin. The oceanic ridges are the site of numerous shallow foci Earthquake (occurring within 70 km of Earth's surface) as well as volcanoes and volcanic islands such as Azores, Ascension, Iceland and Tristan da Cunha.

Inactive aseismic ridges and rises are devoid of any Earthquake activity or volcanic activity. They are mostly arranged in the form of a linear chain of extinct volcanoes and are indicators of plate movement in the past. The most important aseismic systems are the Galpagos rise, Rio Grande rise, Walvis rise and the Hawaiian island chain.

Austral-Gilbert-Marshall chain and Tuamotu were formed when the tholeiitic magma ascending from the hot spot in the form of mantle plume broke through the moving plate forming a volcanic cone which became inactive as the plate moved and the magma source was cut off. In this manner, inactive cones were linearly arranged (also showing direction of plate movement) but the end that is still over the hot spot continues to be active. The system of ridges and rises separate the ocean basins into a series of sub basins. The deep waters in these basins are isolated from each other by the height of the ridges separating them. A low ridge allows good mixing of deep water between basins while a high, continuous ridge effectively isolates deep water. An example of this is the Angola Basin, located in the bight of west coast of Africa. This is cut off from the Brazil basin in the west by the Mid-Atlantic ridge and from Cape Basin to the south by the Walvis Ridge.

Ocean Trenches: Ocean trenches are produced by the subduction of oceanic crust under oceanic or continental crust; therefore, the ocean trenches have parallel island arc or young mountain belts on their seaward side. The most important examples of trenches are the Mariana trench of Philippines, the Java trench, the Aleutian trench, the Japan trench, the Puerto Rico trench, the Peru-Chile trench, Caribbean trench and the Kurile Kamchatka trench.

The system of oceanic trenches girdles the entire Pacific Ocean marking the subduction of the Pacific ocean in the east, northeast and north, and of the Nazca plate in the southwest. In the east Pacific, double trenches are found. They are formed by the abandonment of the old subduction zone and formation of a new one in its place as a result of variations in stress in the lithosphere.

Island Arcs: Island arcs form when two oceanic plates converge and collide. The convergence forces one of the plates having higher density into the mantle. As the plate descends, melting takes place. This melted portion rises upwards and piles up on the adjacent oceanic plate. The constant extrusion of magma from the feeder pipe forms a volcanic cone emerging above the sea level. These islands are arranged in an arc shaped manner.

Each island arc is generally made of two parallel arcs of islands 50-150 km. apart. The outer arc is dominantly composed of folded and thrust sediments and lacks volcanic activity while the inner arc is composed of volcanic islands. The outer arc may have different forms, it may be submerged (south of Java), it may be poorly developed (Antilles and Barbados), it may be dominant (Ryukyu islands between Japan and Taiwan), or it may be topographically and structurally fused with the inner arc (Japan).

Marginal Ocean Basins: Most of the marginal ocean basins are located in areas where crustal deformation or mountain building is active. Eventually, they may change or get incorporated into the adjacent continental block. There can be three types of ocean basins.

Ocean basins can be associated with island arcs and submarine volcanic ridges. For example, many small basins around Indonesia and along the Pacific margin of Asia were formed due to the growth of island arcs which isolated segments of oceanic crust to form small independent oceans. These basins are being extended by back arc spreading (extension and spreading of the seafloor behind the island arc) as a result of complex convective eddies in the asthenosphere above the subducting plate or by pulling away of the adjacent plate. This creates regional tension which results in the formation of a back arc basin characterised by crustal thinning and block faulting.

Some ocean basin exist as marginal sea between two continental blocks, e.g., Mediterranean Sea between Europe and Africa, Gulf of Mexico and Caribbean Sea between North and South America. The Mediterranean and the Blacks Sea are basins formed by the closing of ocean basins as a result of convergence between Africa and Asia. These seas are remnant basins of the once vast Tethys

Sea. As Africa and India moved northward, the Tethys Sea gradually closed at some places completely, as between India and Eurasia, or only partially as between Africa and Eurasia. If the present plate motion continues, the Mediterranean basin will be eventually be enclosed and African plate will be thrust under Alps as Indian plate is thrust under the Himalayas.

Long narrow marginal seas are the third type of ocean basins, such as Red Sea and the Gulf of California. These have formed due to continental rifting in the initial rift zone where the continents split and began to spread apart. Basalt from the upwelling mantle commonly fills the new opening during the early stages of rifting and sediment eroded from the adjacent continents fills the ocean basin. As the sea floor continues to spread, the intervening sea develops the characteristics of a major ocean basin having an oceanic ridge, abyssal plains and continental slope. The Arctic ocean is in this stage of transition from sea to ocean.

Ocean Plateau: Ocean Plateaus are submarine elevations of considerable extent with relatively flat tops. They rise upto 1 or 2 km but do not reach the ocean surface. They are made of thicker more buoyant material, which are not characteristic of the ocean floor. These plateaus result from a variety of processes—some are stretched continental crust, for example, Rockfall; some are produced by the addition of magma to the crust in the middle of plates like Hawaii; and some are produced at ridges, for example, Iceland.

Plate Tectonics and Wilson Cycle: The various features found in the ocean basins are not permanent, as the ocean basin itself is not permanent. For example, the Pacific Ocean is closing because the Pacific plate is subducting very fast beneath the Eurasian plate. The ocean basins emerge from the fragmentation of two continents, enlarge and expand (sea floor spreading) and eventually close. This process of opening and closing of the ocean basin is called formation of Supercontinent or Wilson Cycle.

Earth has been cooling since its formation. Since the crust formed about 3000 million years ago, Earth has undergone seven, or may be eight cycles of Supercontinent formation. Supercontinents break up, forming a new ocean, which grow and later disappear, forming another Supercontinent. The cycle, from start to finish, takes about 500 million years.

Continental crust acts like a blanket and does not conduct heat as effectively as oceanic crust. Thus, a continent remaining in one location for a long time leads to heating of the underlying mantle since it retards heat flow.

As the mantle warms, it expands, elevating the overlying continent and stretching its crust. Eventually, the crust breaks, forming a rift valley. This is what is happening in Africa, which has been in its present location for about 200 million years and stands about 400 m higher than other continents. As the rift grows, a long narrow ocean basin forms. Red Sea is one such example, which is becoming a narrow dean basin. As the rift enlarges and divergence takes place, the narrow ocean basin, bordered by continental margin, widens and along the line where rift developed magma comes from below and solidifies forming an ocean ridge. As more and more material is added on to the ocean ridges by upwelling from below, the ocean floor spreads and the ocean basin widens. To accommodate the widening ocean basin, subduction occurs.

Through time, the oldest parts of the lithosphere in the ocean basin become dense enough to sink in the mantle. As the ocean basin ages, subduction becomes more widespread. Eventually, the ocean basin begins to close as subduction overwhelms spreading due to the mid-ocean ridge. In the initial stages of the ocean basin closing, subduction of one plate causes the development of island arcs and trenches around ocean basins. The Pacific Ocean represents this stage. As subduction proceeds, the ocean basin narrows down forming irregular seas with young mountains. The Mediterranean represents this stage. It is in its terminal stage of closing. Finally, the ocean basin is completely enclosed and young to mature mountain belts, which were on either side of the basin, are sutured to form one single continent. Himalayas, which are the sutured site by the complete closure of the Tethys, represent this stage.

Sea Floor Spreading

It is now fair to say that almost all Earth scientists are convinced that continental drift has occurred and of all evidences no particular piece of evidence is more convincing and crucial than that from palaeomagnetism. But while palaeomagnetic directions have been instrumental in proving continental drift, they also form the critical evidence in favour of another hypothesis—sea floor spreading.

During the fifties and sixties, technological developments permitted extensive and detailed mapping of the ocean floor. One of the most important discoveries was the worldwide Mid Oceanic Ridge (MOR) system. This is a rise or swell with its crest 23 km higher than abyssal plains and a rift valley occupying its crest for the entire length suggesting that great tensional forces are at work. In addition, high heat flow and volcanism was also found to characterise the ridge system. Additional discoveries were those of flat topped seamounts (guyots) which showed signs of being former volcanic islands and dredging of the ocean floor showed that the ocean floor was not older than the Mesozoic, i.e., the ocean floor is relatively young.

In 1960, H.H.Hess came out with a radically new interpretation of MOR system. He postulated that ocean ridges are situated over rising limbs of convection currents in the Earth's mantle. As rising material from the mantle ascends, it spreads in two opposite directions after meeting the lithosphere and creating a tension, which ultimately gets rifted. This rift provides pathways for magma to come to the surface in form of volcanic activity, which solidifies and forms new crust. Areas in which new crust is formed along rising magma are called spreading centres. The generation of new crust, in turn, pushes the crust in two opposite directions. As the convection currents spreads laterally, they carry the seafloor along with them. This lateral movement of the crust is called sea floor spreading. Beneath the deep-ocean trenches, limbs of mantle-wide convection descend. Areas of descending crustal material are subduction zones. Here, the oceanic crust is plunged into mantle and gets destroyed. Thus, whatever new crust was generated did not led to an increase in the surface area, as it was compensated by the melting and destruction of the ocean floor. So, the ocean floor crust is being constantly renewed and thus, it is much younger. No doubt, rocks, older than Mesozoic, are not found. In this process, there is opening of ocean basins, as the ocean floor is carried by convection currents—this process came to be known as sea floor spreading

The evidence, that the sea floor actually spreads, came from palaeomagnetic studies. That, rocks acted like magnets, was abundantly clear and that they helped in delimiting the direction of Poles as well as polarity (North or South) was also very much clear. Polarity such as that existing today is referred to as normal while opposite polarity as reversed. During bathymetric survey of the Pacific Ocean, it was found that some rocks of the ocean floor pointed towards north and some south, thus, revealing a roughly N-S striped pattern of magnetisation. Normally, rocks should be pointing towards north, as this was the direction they acquired while cooling but how does one explain the magnetic reversal in the direction of rocks pointing towards the south. Initially, it was thought that this phenomena of reversed magnetism could be produced either due to an alternating reversal of the Earth's polarity (field reversal) or due to certain properties within the rock itself, as might be brought about by physical and chemical changes in the magnetic minerals during or soon after rock formation (self reversal). There are several evidences in favour of field reversal—

1. Rocks of different types—igneous, metamorphic and sedimentary have the same direction of magnetisation showing that magnetic reversal could not possibly be due to internal chemical and physical changes among the rocks themselves.
2. The Permian, Pleistocene and Late Pliocene rocks show magnetic reversals all over the globe. Had there been self reversal there would be no time correlation. A rock with a self reversing property is likely to occur at any time and if all the rocks of a given age are collected together some should be found to be reversely magnetised while others are normally magnetised.

Self reversal is possible under laboratory conditions and is also a natural process exhibited by some igneous rocks but the evidences are overriding in favour of field reversal. A large number of polarity records are found throughout the geological record. During the last 3.6 m.y., there have been two periods with normal polarity as of today and two with reversed polarity. These main periods are known as geomagnetic polarity epoch and they are named after investigators of various nationalities who have made important contributions to the knowledge of the Earth's magnetism. Gilbert, Gauss and Matuyama Brunhes are examples of these synonymous epochs. Within these epochs are shorter polarity events to which place names have been given commemorising the localities where they were first recognised. From the sequence of magnetic anomalies and their radiometric ages, a reliable chronology of magnetic reversals has been established for the last 4 m.y. In addition, extrapolation as far back as 76 m.y. reveals a sequence of at least 171 reversals.

It can now be inferred that the rocks did not change their polarity, it is the Earth's magnetic Pole that reversed its polarity, i.e., the north magnetic Pole became south magnetic Pole and south magnetic Pole became north magnetic Pole. This is very much possible and such change can be brought about by reversal of the convection currents in the outer core which is apparently responsible for Earth's magnetic field.

The mystery concerning the patterns of magnetic readings with stripes of alternating positive and negative anomalies found on ocean ridges can now be explained. As fractures form on the ocean floor, basalt is intruded into it and is solidified. As it solidified it was magnetised in the direction of the (then) existing magnetic field and thus basalt extruding along the oceanic ridge formed a zone of new crust with normal polarity. With new intrusion later on, this zone of crust split and migrated away from the ridge but remaining parallel to it. When the new intrusion solidified, it also aligned itself in the direction of the Earth's 'then' magnetic field. But by then, the polarity had reversed and thus the new crust generated at the oceanic ridge was magnetised in the opposite direction. In this way, the sequence of polarity reversals became imprinted as magnetic stripes on the oceanic crust.

It is notable that the pattern of magnetic stripes on the ocean floor on either side of the ridge matched the patterns found in a sequence of recent basalts on the continents. The crest of the ridge shows normal polarity and is flanked on either side by a broad stripe of rocks with reversed polarity (formed during the reversed epoch), then follows a stripe of normal polarity containing one narrow band with reversed polarity and so on. In brief, the patterns of magnetic reversals away from the crest of the ridge are the same as those found in vertical sequence of rocks on the continents from youngest to oldest. The striped pattern should, therefore be, bilaterally symmetrical on either side of the median rift. This was what Vine and Matthews suggested. Thus, they converted the 'conveyor belt' type of ocean floor of Hess into a 'magnetic tape recorder'.

Age of oceanic crust	Become progressively older rocks away from ridges, particularly well exemplified by accessible rocks of oceanic islands.
Age of ocean floor	Become progressively older away.
Sediments	From ridges; repeats and confirms pattern of age symmetry displayed by crustal material.
Magnetic polarity	Variation in Earth's magnetic field arranged in bands parallel to each other and to ridge axis.
Magnetic polarity reversals	Patterns of normal and reversed polarity, as preserved in the palaeo-magnetism of oceanic crustal rocks; show bilateral symmetry about ridges.
Continental margins	Topographic and strati-graphic congruence.

While the origin of magnetic stripes was explained, it was also clear that as the new ocean floor was created, it led to the spreading of ocean floor. The distance between the magnetic stripes enabled the rates of ocean floor spreading to be calculated. Same anomalies do not occur at the same distances from the different ridges, and hence, spreading rates must have varied from area to area such as in Equatorial Pacific where the widths of magnetic stripes are wider for corresponding time intervals than those of Atlantic Ocean. Hence, it can be concluded that a faster spreading rate exists for the spreading centre of the Pacific as compared to Atlantic.

It is known that spreading can be recorded in the form of linear magnetic anomalies. If spreading had been continuous about all ridges, the same sequence of anomalies should be reproduced across all ridge-flanks and indeed the same sequence of anomalies is observed above ridge-flanks in all the ocean basins. The worldwide correlation and symmetry of magnetic anomalies provide compelling evidence for sea floor spreading.

Although, sea floor spreading can be observed directly only with great expense and difficulty at sea, in Iceland many of the processes can be seen on land. Iceland is the only large island lying across a mid oceanic ridge and here sea floor spreading can be measured. Spreading in Iceland occurs at rates similar to those found at the crest of mid oceanic ridge. Northeast Iceland has been quiet for 100 years when volcanic activity in 1975 made a 80 km long 5 km wide rift in six years.

It is also possible to check the rate of plate motion by using distance between seamounts in conjunction with radiometric dating. For example, the distance between the islands of Midway and Hawaii is 2700 km. Midway was an active volcano 25 m.y ago when it was located above the hot spot, currently occupied by Hawaii. In other words, Midway has moved 2700 km in 25 m.y. or 1 km per year.

The Pacific, Nazca, Cocos and Indian plates are moving at higher rate than the African, American, Eurasian and Antarctica plates.

The Evidences

1. The crest of the oceanic ridge is characterised by an abnormally high heat flow (9.5 HFU in contrast to 1.2 HFU for the entire ocean floor) and volcanic activity, implying the generation of new crust. Addition of new material leads to shoving and consequent spreading.
2. Linear Magnetic Anomaly patterns. Oceanic crust is characterised by linear to curvilinear magnetic stripes that

roughly parallel the crests of the ridges. The study of these magnetic stripes reveal that they are alternatively normal and reverse to the direction of the existing magnetic fields of the Earth. The explanation lies in the fact that as magma is emplaced along the crest of the ridges, the magnetically susceptible minerals are aligned in the direction of the existing magnetic field of the Earth. It has been noticed that the magnetic field reverses its direction every 10 years. When normally magnetised stripe diminishes the Earth's field, negative magnetic anomaly is produced. Thus, each stripe implies a phase of lava emplacement and with successive addition of lavas to the oceanic crust along the crest of the ridges, the oceanic crust has been spreading.

3. Age of the oceanic floor. The radiometric dating of the ocean floor suggests that rocks get progressively older as one goes away from the ridge crest. This is consistent with the fact that with each fresh emplacement of magma along the ridge crests, the older as one goes away from the ridge crest. This is consistent with the fact that with each fresh emplacement of magma along the ridge crest, the older rocks are pushed away. The islands, seamounts and guyots exhibit the same trend of age variation. The youngest features occur close to the ridge and the oldest in the proximity of the oceanic trenches. Some seamounts and guyots have not only sunk but also tilted, e.g., Capricorn guyot outside the Tonga deep. It is evident that ocean floor is created at the ridge crest and is consumed at the oceanic trenches. Thus, there is a complete recycling of oceanic floor, which is reflected in the young ages of the present oceans. The present oceans are not older than lower Cretaceous or at the most middle Jurassic.
4. Transform faults. All the mid oceanic ridges are characterised by the presence of transform faults, which tear apart the ridges. These are in fact strike-slip-faults but are designated as transform faults, since they transform the spreading motion of the ridge into underthrusting motion in the trenches. The offsets in these faults are original and do not change with time, rather the activities of the faults are renewed as fresh materials are injected into the sea floor along the ridge crests.

5. The average thickness of sediments on the ocean floor is also abnormally small. Had the ocean been as old as the continental shields, the thickness of the sediments on their floor would have been enormous.

Wilson Theory: J. Tujo Wilson, at the same time provided insights into the nature of transform faults. Wilson suggested that these large fractures connected the global active belts into a continuous network that divides the Earth's outer shell into several rigid plates. Wilson was the first to suggest that Earth was made of individual plates and transform faults were the zones along which the relative motion between Plates was possible. It is transform fault, which provide means by which oceanic crust, created at the ridge crests, can be transported to its site of destruction, the subduction zones, so that the Earth's surface area remains same.

Although, it was proved that the sea floor is spreading, the mechanism was somewhat lacking. Our current ideas on the nature and cause of spreading process have been derived from a better understanding and more detailed consideration of geometry of spreading and seismicity of the Earth. In 1960's the World Wide Standardised Seismographic Network (WWSSN) was set up. The distribution of earthquake epicentre, along ocean ridge crests and focal mechanism deduced for some of the larger earthquakes, occurring on the transverse fracture, confirmed the sense of movement on faults. When the epicentres from WWSSN studies were plotted on a world map it was seen that the seismicity of the Earth defines three structural elements of sea floor spreading model (ridge crest, trench and transform fault) with much greater precision. Moreover, the maximum depths at which earthquakes occur, beneath trench and mountain systems, were seen to be very much greater (up to 700 km) than those beneath ridge crest (up to 20 km). These narrow seismic zones outline essentially seismic areas of the Earth's crust, which appear to behave as quasi-rigid plates. They do not exhibit major internal deformation. Morgan, while exploring the geometry of sea floor spreading, came to the conclusion that much of Earth's seismicity is clearly related to relative movements between quasi-rigid plates and consequently, the new concept came to be known as Plate Tectonics.

Theory of Pagaea

. Pangaea (pan = all + gaea = earth) is a hypothetical continent from which the present continents originated by drift from the Mesozoic era to the present. Wegener hypothesised that the super continent of Pangaea broke up to form (i) Laurasia (North America, Greenland, and all of Eurasia north of Indian subcontinent), and (ii) Gondwana land (South America, Africa, Madagascar, India, Arabia, Malaysia, East Indies, Australia, and Antarctica). These two blocks were separated by a long shallow inland sea called the Tethys Sea.

According to Wegener, the Pangaea was surrounded on all sides by an extensive water mass called the 'Panthalassa' (pan = all + thalassa= oceans) or the primeval Pacific Ocean. Wegener also assumed that in the Carboniferous period the South Pole was near Natal (South African coast) and the North Pole was in the Pacific Ocean.

According to Wegener, the Pangaea started breaking up in the Carboniferous period (about 300 million years back), and the present shapes and relative positions of the continents are the result of fragmentation of Pangaea by rifting and drifting apart of the broken land masses. He was also of the opinion that the continents are made of sial and the ocean basins are made of sima. Sial being lighter is floating over the denser sima.

Directions of Drift

In the opinion of Wegener, the continents drifted in two directions: towards the equator, and towards the west. On account of the equator ward drift, Africa and Eurasia were pushed closer together and the Tethys Sea deposits, located in between the two, were raised up in the form of mighty folded mountains of Alps, Altais, Tien Shan, Zagros, Hindu Kush, Himalayas, etc. The peninsula of India and Africa were separated from the Antarctica and Australia because of the equatorward drift. The reason given for this movement was the gravitational attraction exerted by the earth's equatorial bulge.

The other movement of the earth was towards the west. This westwards movement, in his opinion, was due to the tidal force of the moon and the sun on the continents. On account of this westward drift, North America and South America got separated from Europe and Africa and the Atlantic Ocean came into existence.

In addition to this, Wegener attempted to explain the problem of formation of fold mountains with the help of drift theory. As stated above, on account of movement of Gondwana land towards the equator, i.e., on account of their coming closer together, and the southern land mass (Gondwana land) pressing against the northern, the Alpine and the Himalayan ranges were formed in the Tertiary period in the region occupied by the Tethys Sea. Similarly, the Rockies and the Andes were formed as a result of the westward drift of the Americas and the resistance offered by the rocks of the ocean floor to this movement. Between the two immense rafts, a trail of fragments was left behind to form the islands of the West Indies.

Wegener assumed that the disruption of the southern block of the Pangaea (Gondwana land) took place mostly during the Mesozoic era, while the westward drift of North America occurred during the Tertiary period.

Different Helping Evidences

In support of his theory of continental drift, Wegener gathered geomorphological, climatological and palaeontological evidences. Some of the important evidences given in support of the theory are as under:

The Geographical Similarity in the Opposing Coasts of the Atlantic Ocean: The outlines of the coasts on the two sides of the Atlantic Ocean are such that they can be easily joined together, and one appears to be the detached portion of the other. The eastern coast of South America can be fitted against the western coast of Africa. This was called the *'jigsaw-fit'* of the opposing coasts of the Atlantic Ocean by Wegener.

Palaeontological Evidence: The striking similarity of certain fossils found on the continents on both sides of the Atlantic Ocean is difficult to explain unless the continents were once connected. The fossil record indicates that a new species appears at one point and disperses outward from there. Floating and swimming organisms could migrate in the ocean, from the shore of one continent to another, but the Atlantic Ocean would present an insurmountable obstacle for the migration of land dwelling animals, such as reptiles and insects, and certain land plants.

Fossils of glossopteris, a fernlike plant, have been found in rocks of the same age from South America, South Africa, India and Australia, and within 480 km of the South Pole, in Antarctica. Mature seeds of this plant were several millimetres in diameter, too large to have been dispersed across the ocean by winds. The simultaneous presence of glossopteris on all of the southern continents, therefore, is strong supporting evidence that the continents were once connected.

The distribution of Palaeozoic and Mesozoic reptiles provides similar evidence, as fossils of several species have been found in the now-separated southern continents. An example is mammal-like reptile belonging to the genus *lystrosaurus.* This creature was strictly a land dweller. Its fossils are found in abundance in South Africa, South America, Asia and Antarctica. This genus thus inhabited all of the southern continents during the same geologic period. Clearly, these reptiles could not have swum thousands of kilometres across the Atlantic and Antarctic oceans, so some previous connection of the continents must be postulated. A former land bridge, similar to present day central America, would explain the presence of lystrosaurus in distant parts of the world, but surveys of the ocean floor show no evidence of such a submerged land bridge.

Evidence from Structure and Rock Type: A number of geologic features end abruptly at the coast of one continent and reappear on the facing continent across the Atlantic Ocean.

The folded mountain ranges at the Cape of Good Hope, at the southern tip of Africa, trend from east to west and terminate sharply at the coast. An equivalent structure of the same age and style of deformation appears near Buenos Aires, Argentina.

The folded Appalachian mountains are another example of similar geologic features. The deformed structure of the mountain belt extends northward across the eastern United States and through New Found land, terminating abruptly at the ocean. They reappear at the coasts of Ireland and Brittany.

Other examples could be cited, but the important point is that the continents on both sides of the Atlantic fit together, not only in outline, but in rock types, basalt plateaus and structures. They are related much like matching pieces of a torn newspaper.

The geologic similarities on opposite sides of the Atlantic Ocean are found only in rocks older than the Cretaceous period, which began about 144 million years ago. The continents are believed to have split, and began drifting apart in the Jurassic period, about 200 million years ago.

Evidence from Glaciation: During the latter part of the Palaeozoic era (about 300 million years ago), glaciers covered large portion of the continents in the Southern Hemisphere. The deposits left by these ancient glaciers can be readily recognised, and striations and grooves on the underlying rock show the direction in which the ice moved. Except for Antarctica, all the continents in the Southern Hemisphere show no trace of glaciation during this time. In fact, fossils of plants indicate a tropical climate in that area. This evidence is difficult to explain in the context of fixed continents because the climatic belts are determined by latitude.

Even more difficult is to explain the direction in which the glaciers moved. Regional mapping of striation and grooves indicates that in South America, India and Australia, the ice moved inland from the oceans. Such movement would be impossible unless there were a land mass where the oceans now exist.

Evidence from Other Palaeoclimatic Records: Other evidence of striking climatic changes recorded in the stratigraphic record tends to support the drift theory. Great coal deposits in Antarctica show that abundant plant life once flourished on that continent, now covered with ice more than two kilometre thick.

On the other continents, salt deposits, formations of windblown sandstone, and coral reefs provide additional clues that permit us to reconstruct the climatic zones of the past. The palaeoclimatic patterns are baffling with the continents in their present positions, but if they are grouped together in their present positions, the patterns are easily explained.

Palaeomagnetic Evidence: According to Wegener, there has been change in the positions of the poles in the different geological periods, which he called *'polar wandering'*. The palaeomagnetic data reliably indicate the existence of Pangaea at the end of the Palaeozoic era.

The above evidences for the theory of continental drift were considered and debated for years. Wegener was criticised for failing to explain what forces would permit continents to plough through ocean of rock. The idea of moving lithosphere was yet to come.

Wegener was dismissed as a crank. His critiques claimed, with some justification, that he had carefully selected only those data supporting his hypothesis, ignoring contrary evidence. Where, for instance, were the 'wakes' or 'tracks' through old sea bed that the migrating continents would leave? By 1926, the 'drifters' were in full retreat. When Wegener died on an expedition across Greenland in 1930, his theory was already in eclipse.

Idea of Revival

As modern scientific capabilities built the case for continental drift, the 1950s and 1960s saw a revival of interest in Wegener's concepts, and finally confirmation. Aided by an avalanche of discoveries, the theory today is universally accepted as an accurate model of the way earth's surface evolves, and virtually all earth scientists accept the fact that continental masses move about in dramatic ways now referred as *plate tectonics.*

The concept of continental drift refused to die, however, those neatly fitted continents provided a haunting reminder of Wegener to anyone looking at an Atlantic chart. In 1935, a Japanese scientist, Kiyoo Wadati, speculated that earthquakes and volcanoes near Japan might be associated with continental drift. In 1940, seismologist Hugo Benioff plotted the locations of deep earthquakes at the edges of the Pacific Ocean. His charts revealed the true extent of the *Pacific Ring of Fire*—a circle of violent geologic activity surrounding much of the Pacific Ocean. Seismographs were now beginning to reveal a worldwide pattern of earthquakes and volcanoes. Deep earthquakes did not occur randomly over the earth's surface but were concentrated in zones that extended in lines along the earth's crust. Benioff, Wadati, and others wondered what could cause such an orderly pattern of deep earthquakes. Many of the lines corresponded with a worldwide system of oceanic ridges, the first of which was plotted in 1928 by *Metor* oceanographers working in the middle of the North Atlantic. Notice the odd pattern they form almost as if the earth's lithosphere is divided into sections! Benioff's sensitive -seismographs also began to gather strong evidence for a partially molten, non-rigid layer in the upper mantle. Could the continents somehow be sliding on that?

Other seemingly unrelated bits of information were accumulating. Radio metric dating of sediments and rocks was perfected after World War II. This technique is based on the discovery that unstable, naturally radioactive elements lose particles from their nuclei and change into new, stable elements. The radioactive decay occurs at a constant rate, and measuring the ratio of radioactive to stable atoms in a sample provides its age. To the surprise of many geologists, the maximum age of the ocean floor and its overlying sediments was radio metrically dated to less than 200 million years—only about 4 per cent of the age of earth. The centres of the continents are much older: some parts of the continental crust are more than 3.9 billion years old, about 85 per cent of the age of earth. Why was oceanic crust so young?

Attention quickly turned to the deep-ocean floors, the complex profiles of which were now being revealed by echo sounder—a device that measures depth by bouncing high-frequency sound waves off the bottom. In particular, scientists aboard the Lamont Geological Observatory Deep-Sea Research Vessel, Vema (a converted three-masted schooner) invented deep survey techniques. They probed the bottom with powerful echo sounders and looked beneath sediments with reflected pressure waves. The overall shape of the Mid-Atlantic Ridge was slowly revealed. The ridge's conformance to shorelines on either side of the Atlantic Ocean raised many eyebrows. Ocean floor sediments were shown to be thickest at the edge of the Atlantic and thinnest near this mid-ocean ridge.

Vema and other research ships also compiled more complete, accurate charts of the submerged edges of continents. At Cambridge

University, Sir Edward Bullard used a computer to process these data to achieve the best possible fit of the continental jigsaw-puzzle pieces around the Atlantic Ocean. The fit was astonishingly good.

Mantle studies were keeping pace. The first links in the Worldwide Standardised Seismograph Network, begun during the International Geophysical Year in 1957, were beginning to report data from seismic waves reflected and refracted through the planet's inner layers. This information verified the existence of a layer in the upper mantle that caused a decrease in the velocity of seismic waves. This finding strongly suggested that the layer was not solid. Perhaps, the lithosphere was isostatically balanced in this plastic layer, and continents could move around in it if a suitable power source existed.

Sea Floor Spread

The hypothesis of 'sea-floor spreading' is a new development which proves the theory of continental drift. It was put forward jointly by the late Professor Harry H. Hess of the Princeton University and Robert S. Dietz in 1960.

The idea of sea-floor spreading is a synthesis of the results obtained by marine geologists and geophysicists like Maurice Ewing and his co-researchers. Hess and Dietz proposed sea-floor spreading as the mechanism that builds this mountain chain and drives continental movement. In the opinion of Hess, the submarine mountain ranges, called the mid-ocean ridges, were the direct result of upwelling flows of magma from hot areas in the upper mantle and asthenosphere and perhaps deeper sources. When mantle convection brings magma up to the crust, the crust is fractured and the magma spills out and cools to form new sea floor, building the ridges and spreading laterally.

The marine geologists between 1950 and 1960 had been able to establish the following three important facts about the floors of the oceans:

1. The crust below the ocean floor was found to be only *6 to 7* km thick, whereas below the continental surface this thickness was 30 to 40 km.

2. The existence of the Mid-Atlantic Ridge was known to the geologists and geophysicists, but it was discovered that mid-oceanic ridges were present in all oceans. The ridges are subjected to earthquakes and volcanic eruptions.
3. The rocks of the ocean floors nowhere were found to be older than the Cretaceous period which began about 135 million years ago.

On the basis of the above discoveries, Hess (1960) proposed that the ocean floor was mobile and hot magma (lava) rises up from the earth's mantle by convection currents along the middle of mid-ocean ridges and moves away on either side and finally get lost in the ocean-trenches situated along the continental coasts. It is through this process of spreading that the ocean floor has been built. In other words, the ocean floor is a relatively young feature of the earth's surface. It is constantly being regenerated at the mid-ocean ridges and is subject to continuous lateral spreading until it is destroyed in the trench systems and becomes reincorporated into the mantle. On the other hand, in spite of the fact that the continents are old and apparently permanent features, they drift apart passively and together on the backs of mantle-generated convection currents. These facts prove that the continents and ocean basins are in constant motion.

Hess put forward the following evidences and arguments in support of his theory of sea-floor spreading:

1. The occurrence of earthquakes along the crust of the mid-oceanic ridges.
2. The dearth of sediments at the crests of the mid-oceanic ridges and the active volcanic islands associated with the crests of the Mid-Atlantic Ridge are easily explainable in Hess' theory of sea-floor spreading. Moreover, the thickness of the sedimentary deposits increases with distance from the ridge. This is inconsistent with sea-floor spreading because the more distant parts are older and have had a longer time at their disposal to deposit sediments.
3. The edge of the marine sediments above the basaltic layer goes on increasing with distance from the ridge. For example, in the Atlantic Ocean, the age of the main sediments near

the ridge was found to be less than 7 million years, while the age of sediments farthest away from the ridge has been calculated to be more than 160 million years.

4. There is a reversal in the main magnetic field of the earth known as magnetic dipole.
5. The normal and reverse magnetic anomalies are found in alternate manner on either side of the mid-oceanic ridges.
6. There is parallelism in the time sequence of palaeomagnetic epochs and events calculated for about 4.5 million years on the basis of magnetism of basaltic rocks or sedimentary rocks.

On the basis of the above evidences, it may be inferred that: (i) there is continuous spreading of sea floor, (ii) new basaltic crust is continuously formed along the mid-oceanic ridges, (iii) the newly formed basaltic layer is divided into two equal halves and is thus displaced away from the mid-oceanic ridges, (iv) alternate stripes of positive and negative magnetic anomalies are found on either side of the mid-oceanic ridges. Such magnetic anomalies (positive and negative) are formed because of temporal reversal in the geomagnetic field. The rocks formed during normal geomagnetic field contain positive magnetic anomalies while the rocks formed during reverse polarity (reverse geomagnetic field) denote negative magnetic anomalies.

The sea-floor spreading hypothesis explains a number of other things, such as, (a) continental drift, (b) occurrence of earthquakes near the mid-oceanic ridges, and the absence of sediments on their summits, (c) presence of active volcanic islands on the Mid-Atlantic Ridge, and (d) the absence of rocks older than 135 million years on the ocean floor or the oceanic islands.

The concept of sea-floor spreading is an increasingly convincing evidence about the continental drift. Moreover, the hypothesis of sea-floor spreading has now developed into a more sophisticated and comprehensive theory of plate tectonics, in the context of which the theories of continental drift, and sea-floor spreading appear to be obsolete and outdated.

Plate Tectonics Theory: The term 'plate tectonics' was first used by Tuzo Wilson of the University of Toronto in 1965 but the theory of plate tectonics was first published by W.J. Morgan of the

Princeton University in 1967. This theory is based on the concept of 'sea-floor spreading' advocated by Harry Hess and confirmed by Vine and Mathews by their interpretation of linear magnetic anomalies. It is an improvement over the Wegener's theory of 'continental drift' and has been considered as the most sophisticated and comprehensive theory about the drift of continents and expansion of sea floors.

Plate tectonics is a theory of global dynamics in which the lithosphere is believed to be broken into a series of separate plates that moves in response to convection in the upper mantle. The margins of the plates are sites of considerable geologic activity, such as sea-floor spreading, volcanic eruptions, crustal deformation, mountain building and continental drift.

Tectonics from the Greek *tektonikos,* meaning building or construction, refers to the deformation of the earth's crust as a result of internal forces, which can form various structures in the lithosphere. Tectonics processes include the up-welling of magma, plate movement, subduction of crust, plus folds, faults, warps, fractures, earthquakes, and volcanic activity.

Lithospheric Plates: Plate is a broad segment of the lithosphere (including the rigid upper mantle, plus oceanic and continental crust) that floats on the underlying asthenosphere and moves independently of other plates. La Pichon (1968) divided the earth into six major plates and nine minor plates. These are as under:

Major Plates: 1. African Plate, 2. American Plate, 3. Antarctica Plate, 4. Australian Plate, 5. Eurasian Plate, and 6. Pacific Plate.

Minor Plates: 1. Arabian Plate, 2. Bismark Plate, 3. Caribbean Plate, 4. Carolina Plate, 5. Cocos Plate, 6. Juan de Fuca Plate, 7. Nazca or East Pacific Plate, 8. Philippines Plate, and 9. Scotia Plate.

It is proposed in the theory of plate tectonics that the entire surface of the earth comprises of internally rigid, but relatively thin (100-150 km) plates. These plates are continuously in motion both with respect to each other and the earth's axis of rotation. Virtually all seismicity, vulcanicity and tectonic activity is localised around plate margins and is associated with differential motion between adjacent plates.

Most of the plates include both continental and oceanic crusts. The area of the plates is fairly large in comparison to their depth and thickness. It has also been established that the depth of the plates is even less under the oceanic crust. Three types of motion are possible between the plates: (i) separation or divergent, (ii) closing together or convergent, and (iii) friction or shearing.

Plate Boundaries or Margins: Boundaries of lithospheric plates can be curved as well as straight, and individual plates can pivot as they move. There are many geometric variations in the shapes and motions of individual plates.

Constructive Plate Margins or Divergent Plate Boundaries (Oceanic Ridges): The plates diverge and move along mid-oceanic ridges. Hot material from the deeper mantle wells up to fill the void. Some of this material is erupted at the surface as basaltic lava. Thus new lithosphere is formed at the plate's trailing edge. The oceanic ridges stand high because their material is hot and, therefore, low in density. Heat flow at the ridge crest is six times greater than that of old oceanic crust beyond the flanks of the oceanic ridges. New plates form as old ones break up and drift apart. The ridge represents a zone along which two plates are in motion away from each other, yet they do not separate because new material is continuously added to the near of each. Thus, the mid-oceanic ridges on which new ocean crust is being formed are known as the *'constructive plate margins'*. The rift valleys of East Africa are believed to have formed along a zone of incipient 'divergent plate boundaries' and are characterised by updoming, rifting, and volcanism. A more advanced stage of rifting is exemplified by the Red Sea which almost completely separates the Arabian peninsula from Africa.

The ridges are commonly offset by large faults that are also the sites of shallow earthquakes. It is here that plates slide past one another. The best known example of this type of fault is California's San Andreas fault.

Destructive Plate Margins (Ocean Trenches): The second type of margin occurs at the deep-ocean trenches. At ocean trenches two plates approach each other and one slips down under the margins of the other at an angle of about 45 degrees. Thus plate boundaries at which the net effect of the motions is to destroy surface area are called *'destructive plate margins'*.

In brief, crust is destroyed at convergent plate boundaries—regions of violent geological activity where plates are pushing together. South America, embedded in westward moving South American plate, encounters the Pacific's Nazca plate as it moves eastward. The relatively thick and light continental lithosphere of South America rides up and over the heavy oceanic lithosphere of the Nazca plate which is subducted into the deep trench that parallels the west coast of South America.

The subduction plate's periodic downward lurches cause earthquakes. Some of the oceanic crust and its sediments will melt as the plate plunges downward, forming a magma rich in water and carbon dioxide. In places this magma rises through overlying layers to the surface and causes volcanic eruptions. The active volcanoes of Central America and South America's Andes mountains are product of this activity, as these are the areas of numerous earthquakes. The North American cascade volcanoes, including Mount St. Helens, result from similar processes. Most of the subducted crust mixes with the mantle, some of it eventually reaching into the mantle to depths of at least 400 km.

Many island arcs and peripheral trenches of the western and northern Pacific Ocean result from the convergence of two oceanic plates. Plate convergence (and divergence) is faster in the Pacific Ocean than in the Atlantic Ocean, in a few places reaching a rate of 18 cm (7 inches) a year. This is the source of *'Pacific Ring of Fire'*.

Conservative or Passive Plate Margins: These are the margins at which the plates neither gain nor lose surface area, but simply slip past each other. This happens along the transform faults and thus the crust is neither created nor destroyed. The potential for earthquakes may, however, be great as the plate edges slip past one another. The eastern boundary of the Pacific plate is a long transform fault system. California's San Andreas fault is merely the most famous of many faults marking the junction between the Pacific and North American plates. It is believed that some 50 million years from now, the western California will encounter the Aleutian trench.

Major Assumptions

There are three basic assumptions of the theory of plate tectonics:

1. That sea-floor spreading occurs, i.e., new oceanic crust is continuously generated at irregular line 'sources' (active oceanic ridges).
2. That the area of the earth's surface is fixed and during the last 600 million years the radius of the earth does not appear to have increased by more than 5 per cent. In other words, the area of the crust destroyed almost equals the area of new crust created.
3. That once formed, new crust forms part of a rigid plate, which may or may not incorporate continental material.

Why Plate Motions ?: The main possible causes of plate motions are as under:

Thermal: All active oceanic ridges are characterised by a wide scatter of heat flow values, some of which are very high. With increasing distance from the ridge crest, the scatter diminishes and the mean heat flow falls until it reaches average level for the oceans. On the other hand, oceanic trenches have abnormally low heat flow but a short distance away, in the adjacent island-arc, the flow is high.

Rates of Motion: Spreading occurs symmetrically at ridges at rates ranging from less than 1 to 6 cm per year. At oceanic trenches crust is consumed at rates from 5 to 15 cm per year.

Oceanic Crust Formation: Whatever the detailed process of crustal deformation at mid-oceanic ridges, it seems that crust of approximately the same thickness is generated whatever the spreading rate.

Oceanic Topography: The mechanism must be consistent with the development of topographic ridges at centres of spreading. Ridges rise 2 to 4 km above the sea level of the ocean floor and near the axis slope away more or less symmetrically from the crest.

Gravity: Observations show that ridges are close to isostatic equilibrium. Sinks are characterised by topographic trenches which may to a greater extent be filled with sediment; they are strongly out of equilibrium and show the largest negative gravity anomalies recorded on the earth.

Distance Separating Sinks and Sources: This is highly variable. In some cases newly formed lithosphere may travel only a few hundreds of kilometres before it is consumed. In other cases, the sink may be two or three thousand kilometres apart.

Strength of the Lithosphere: Even large lithospheric plates appear to be able to move great distances without undergoing significant internal deformation. In some cases the plates are twenty times as long as they are thick. With such a length-to-thickness ratio neither compressional nor tensional stresses could be transmitted from one end of the plate to the other unless the frictional resistance beneath the plate was very small. On account of the presence of high heat flow near the ridges several scientists have suggested that the movement of the surface is related to some sort of convection in the interior of the earth. Recently, D.L. Turocotte and E.R. Oxburg have put forward the concept of 'thermal boundary layers' based on thermal convection. On account of thermal instability in the mantle, heat is conducted upward through the movement of molten material. Once this kind of movement has started, the pattern takes the form of convection currents. The thermal energy generated by conventional current is the main cause of movement of the major and minor plates.

There is thus no lack of possible driving mechanism: plates may be pushed apart by mantle upwelling, pulled apart by sinkers beneath the trench systems, carried by convection currents in the atmosphere or slide under the influence of gravity. Moreover, stresses set up between and within plates may in part determine their motions. This may be particularly relevant in the case of small plates. It seems probable that plate movements are not solely attributable to any one of these effects but result from a combination of all acting in varying degrees on individual plates.

Thus, the new geological and geophysical data acquired since 1960s amply confirmed the original hypothesis of 'sea-floor spreading' and led to its extension and gradual modification, particularly with regard to mechanism. At present, there is no serious obstacle to the acceptance of sea-floor spreading and continental drift as facts rather than theories, especially in the light of the results of the deep-sea drilling.

The theory of plate tectonics brought a revolutionary change in the earth science and now the continental drift theory has been proved beyond any doubt with the help of sea-floor spreading and plate tectonics. These theories also explain the present shape and arrangement of the continents and ocean basins as well the occurrence of earthquakes and volcanoes in the weaker zones of the earth crust.

Importance of Volcanicity

Davisian Theory

The first model and also the most accepted model in geomorphology was developed by William Morris Davis under the title of "Cycle of Erosion". During this period, the Darwinian concept of evolution was incorporated in many scientific and social fields. Davis too, incorporated the themes of time and evolution into his concept. Basically, the concept of Davis was built around four assumptions.

Other Theories

1. Following Darwin, he asserted, "Landforms, like organic forms, shall be studied in view of their evolution". Evolution implied an inevitable, continuous and broadly irreversible process of change producing an orderly sequence of landform transformation, wherein earlier forms could be considered as stages in a progression leading to later forms. Time was to be viewed not as a temporal framework within which events could occur, but a process that imprinted progressive changes in landform geometry from which the passage of time could be inferred.
2. The landform's initial uplift is the chief source of energy in the form of potential energy and thereafter, there is an

irreversible equalisation of energy levels throughout the landform assemblage, leading ultimately to a spatially uniform terrain, which Davis called peneplain.

Thus, each stage of the cycle was associated with declining potential energy. As the landform was worn down and each stage was characterised by an assemblage of landforms having geometries appropriate to the local potential energy, expressed by the difference in the level between the land surface and some lower elevation (base level towards which degradation was directed).

3. The uplift of the land is rapid and while the land mass is being uplifted, there is very little or no erosion. Thus, the available relief difference between the uplifted surface and the general base level, or the inland projection of sea level, which forms the general lower limit of fluvial degradation, is considerable. Although, he stated that other types of relief, as well but these types were not much common.
4. Uniform lithology. The landform in question was of homogenous structure and uniform lithology.
5. He incorporated the concept of graded condition, which is first achieved near the mouth of a river spreading headwards through drainage lines and up the associated valley side slopes towards the divides.

Statement: Based on these assumptions, he proposed:

> "Each type of regional geological structure could be viewed as evolving under the protracted operation of a given process to produce a sequence of stages of landform assemblages. Each stage would possess a suite of landforms appropriate to and characteristic of it"

By structure, he meant more than is usually meant today. It not only included the attitude of rocks, the nature of the dip of the beds, their folds and faults, but also what would now be termed the lithology of beds, the nature of the rocks, their relative hardness and relative permeability.

Process includes the different agents of weathering and erosion, such as physical, chemical and biological, various mass movements and work of water, wind and ice.

Stage means the length of time. Davis divided the time during which processes have acted into stages of Youth, Maturity and Old. These were, however, relative time scales and cannot be related to any particular time frame. The use of the names Youth, Mature and Old emphasises the biological origin of his theory.

Landforms: Open and Closed

A closed system is one, which is closed with respect to matter but energy may be transferred between the system and its surrounding. An open system is one in which both matter and energy can cross the boundary of the system and be exchanged with the surroundings. Landform assemblages are open systems. Landforms are open to inputs of potential energy associated with their uplift, of kinetic energy from precipitation, of thermal energy from the Sun, and chemical energy released as a result of the breakdown of rocks at or near the earth's surface.

Landforms are open to export of various types of energy—thermal energy as the result of stream discharge and air movement; of kinetic energy by stream flow and sediment load and chemical energy by dissolved stream load.

A low entropy closed system is one, which possesses differentials in the distribution at energy in such a way that the flow from high to low energy location is capable of performing work. With time, the energy within the system becomes more equally distributed, the entropy increases till maximum entropy is reached. When all parts of the closed system have the same energy levels, no flow of energy is taking place and no work is being done in the system. Landforms do no behave in this way, and hence, are not closed system.

Process of Grading

In adjusting to changes in discharge and erodibility of its bed, a stream modifies its channel so that irregularities are minimised and the least energy is expended in the movement of water and sediment along its course. Frequently this involves enlargement of the channel cross section by erosion, or reductions of its dimensions through deposition.

The overall tendency is towards a smooth long profile—a profile of equilibrium in which all factors are in a state of balance Geologists have long used the term grade, in referring to a stream that has achieved such a balance. A graded stream is one in which the slope has become so adjusted, under conditions of available discharge and prevailing channel characteristics, that the stream is just able to transport the sediment load available to it. If any of the controlling factors change, then the stream will adjust to absorb the change and restore an equilibrium position.

Local Base Level: In actuality, it is unlikely that a condition of perfect equilibrium is ever achieved in natural stream systems. In the drainage basin of a typical river changes are constantly taking place that upset the balance. A heavy rain may suddenly increase the discharge in one tributary, collapse of a bank may locally introduce an excess of sediment to the stream, or the stream may suddenly encounter a less erodible rock along its course.

Thus, the graded stream is a system in equilibrium. Its diagnostic characteristic is that any change in any one of the controlling factors will cause a displacement of the equilibrium in a direction that will tend to absorb the effects of change. Davis suggested that a smooth, concave long profile (a graded curve) was the manifestation of a state of equilibrium in a river which was then said to be graded river. It is now known that to produce a state of grade it is not necessary to have a smooth profile stage of its development not simply in the later stages as suggested by Davis.

Youth Stage

1. A subaerial or submarine surface of low relief is uplifted producing a region of broad, poorly defined stream divides. These stream divides are separated by a number of trunk rivers, some large tributaries and many short tributaries, which are engaged in headward erosion.
2. This headward erosion and vertical incision of the whole drainage network goes on to increase relief throughout youth. The result is the creation of steep V-shaped valleys flanking stream courses. These are initially irregular as minor structural controls produce falls and rapids.

3. Weathering of the summit areas produces a mantle of rock waste, whose wholesale slipping causes the minor structural controls (which have helped in creation of falls, rapids and temporary lakes) to be smoothened.
4. Thus, the initial lakes are drained and removed as the rivers keep cutting vertically. This happens at the end of early youth.
5. By the end of middle youth, falls and rapids are removed by erosion in the trunk streams.
6. The original flat summit areas are few in number and a larger part of the landscape is dominated by steep valley sides where the surface weathered debris are close to the angle of repose.
7. As weathering proceeds, the waste mantle derived from weathering covers all irregularities on the surface and the angle of repose decreases.
8. This weathered product is transported by rivers.
9. As the rivers carry this load, their ability to erode becomes less because of the transfer of energy in transporting the products, and thus, deposition starts.
10. Finally, a balance is reached between erosion and deposition, i.e., there is an equilibrium between erosion and deposition, so that velocity of water is just enough to transport the sediment load supplied from the drainage basin and neither erosion nor deposition takes place. This is a graded condition, which is attained by major streams by the end of late youth and is evidenced by the smooth unbroken curves.
11. Grading of the main stream also means that small floodplains form in the lower part of the main river.

Stage of Maturity

1. By the beginning of maturity, the region has the maximum available relief (since by this time vertical incision has stopped and lateral cutting has started), the drainage network ceases to increase (as all the streams are fully developed by now), and thus, the area is covered with well integrated drainage network which has maximum relationship to bedrock geological structure.

2. The main stream and their main tributaries are graded by now. The rivers have started forming floodplains.
3. The graded condition spreads headward up the tributaries as well as valley side slopes, which begin to decline in angle, after the early maturity maximum.
4. The divides become narrower, sharper and ridge like and the slopes decline in angle.
5. The floodplains are produced as a result of lateral erosion. Meandering occurs along major rivers but the flood plains in general do not exceed the width of meander belts.
6. The relief decreases progressively as the summits are lowered more rapidly while the valley floors are lowered exceedingly slowly by the graded rivers.
7. Not only the main streams but even the tributaries and wet weather rills are graded.
8. The whole land surface is covered by a graded waste mantle and slopes become less steep due to communition of the debris forming the waste mantle.
9. So, by late maturity, the topography is composed almost of valley side slopes and flood plains to a lesser extent.

Stage of Old Age

1. The slopes by now are covered by a thick waste mantle, which prevents any further weathering of the rock below. Moreover, this waste mantle covers any lithological difference which were important in the formation of relief in youth.
2. The debris in the waste sheet becomes finer, and consequently, the angle of the slope of the waste becomes less and less.
3. The valley sides, slopes and divide crests are graded and the landscape is composed of broad open and gently sloping valleys containing broad flood plains many times the width of associated meander belt, surmounted by very slowly lowering rounded divides.
4. The whole surface approaches closer and closer to the base level under a thickening cover of creeping soil and the relations between terrain and geology become masked. A peneplain is thus formed.

The Criticisms

1. The cycle requires a long time and for the entire time, the conditions are hardly stable. Interferences, such as climatic changes or alterations in the sea level are bound to upset the orderly progress of the cycle. Climate is at no time sufficiently stable over prolonged periods, there are glacial and interglacial periods. Pleistocene glaciation was one such period, which was not taken account of.
2. Cycle concept is too theoretical, landscapes behave in a much more complex manner than Davis envisaged.
3. Davis did not explicitly define 'normal' but implicitly he meant the assemblage of processes dominating the temperate landscapes of North America and Europe where he worked, and especially the action of running water. Any landscape can be considered normal with which geomorphologists are familiar. Moreover, in view of the proportionately small extent of the humid temperate mid-latitude landscapes, the application of 'normal' concept is questionable. Again, in the context of conditions that have occurred in the geological past, present conditions are not normal, they are unique.
4. The initial assumption of rapid uplift is unsatisfactory. It is rarely that landforms experience rapid upliftment, although it must be said that some areas experience extremely high rates of uplift and mean rates of orogeny could well be about eight times greater than the maximum rates of denudation.
5. Landforms do not present smooth surfaces for the commencement of the cycle.
6. Although, Davis said that landform is a function of structure, process and stage, he clearly over emphasised stage and also structure to a lesser extent.
7. It is doubtful whether youth-maturity-old age sequence and the landforms associated with each stage occur. The only proof of the cycle is its own logic and internal cohesion. The theory is not testable, except against its own assumptions and conclusions. The reality is that, if any landscape with features of, say, maturity is observed, it cannot be said that

it is mature because it cannot be proved that it has been derived from a youthful landscape. For example, the Gangetic plain presents many diagnostic features of mature landscape but it has not been derived from a youthful landscape. At least four points can be noted in this regard—

(i) Some landscapes have features at both mature and old age.

(ii) Some landscapes do not fit into any of the categories—youth, mature and old. Landscapes exhibit an infinite gradation of form.

(iii) It is doubtful whether surfaces flat enough to be called peneplains can be ever created by slope decline as the capacity for work in the landscape associated with declining potential energy becomes very low.

(iv) There is no evidence that slope evolves the way Davis suggested. As a model of slope form and evolution, the cycle has proved very misleading.

8. The concept of grade has proved elusive. It is difficult to envisage graded rivers and graded hillslopes coexisting with long term degrading of the landscape.

9. Davis in his study omitted the mechanics and nature of present day processes in both field and theory. This reflects the descriptive and non-quantitative nature of Davis's work. He completely neglected biological processes. Davis's landscapes looked like deserts. This is surprising because Davis based his work on the well vegetated temperate mid latitudes, and this neglect meant that his title 'geographical' was also inappropriate.

10. It is based on deductive approaches, i.e., arguments on certain assumptions and proceedings or inferring from the general to the particular. This is unscientific. Scientific method is inductive approach, which builds on experimental evidence on particular instance to produce general inferences. No doubt, it was the source of many errors. For example, meandering is not only associated with maturity and old age, it can also occur in youthful stage.

The Analysis

Penck's main emphasis was on the development of slopes in the evolution of landforms. However, the development of slopes is far simple involving many complexities.

Theory of King: L.C King's model of landscape evolution is similar to Davis' in that it assumes that uplift is episodic and rapid in comparison with rates of denudation, and that the overall morphology of a landscape at any point in time is diagnostic of its evolutionary stage of development. However, King's scheme is significantly different from Davis. The difference lies in the mode of slope development he proposed. King initially developed his model to account for the landscapes of southern Africa. These are characterised by extensive, gently inclined surfaces dotted with inselbergs and separated by escarpments, and have developed under predominantly arid to tropical wet-dry morphoclimatic regimes.

Rather than the sequential replacement of parallel retreating slope segments by lower angle elements, King envisaged the parallel retreat of a single free-face slope unit, leaving a broad, concave pediment sloping at an angle of 6-7° or less at its base. Gradually over time, pediments coalesce to form pediplains and this mode of landscape development, is therefore, called pediplanation.

King theorised that once pediment surfaces have been formed, they persist with little change until the next phase of surface uplift promotes a new cycle of river incision and escarpment retreat, which consumes existing pediplains and creates new ones. As in the Davisian model, the dating of such denudational episodes can be described in terms of the timing of the fall in base level initiating each new landscape cycle. None the less, the land surface itself is diachronous because in King's model landscapes essentially develop through backwearing as escarpments experience parallel retreat; land surfaces, therefore, are progressively older away from escarpments. Consequently, it is possible to talk of the local age of land surface and even to refer to a terminal age determined by the final removal of a pediplain remnant.

Slope Development: Davis and Penck: An essential corollary to the concept of cycle of erosion is the cycle of slope development leading to peneplanation. During fluvial incision, relatively steep slopes are formed, but as the river reaches what he called a graded profile downcutting ceases and hillslope processes lead to a gradual waning of the valley sides. Davis was never very explicit about the exact nature of these processes, but he showed that the consequence is the development of concavo-convex slopes of progressively lower angle.

Such slope forms are, in fact, common in many humid temperate regions. It is apparent that different hillslope processes tend to operate on different parts of the slope, and together these may result in slope decline. Thus, soil creep is active on the convex crest area. Here, there is no input of material from upslope while rates of transport increase as the slope gets steeper downslope. Thus, soil creep is able to transport the debris as it weathers, little material builds up and over time, this zone undergoes gradual lowering. By contrast, the steeper, rectilinear mid-slope receives inputs of material from upslope as well as from weathering of the underlying bedrock. Larger quantities of material must be transported through this segment, therefore, a rapid mass movements often occur due to periodic oversteepening of the slope and the development of thick, unstable mantles of debris. Over time, this segment is also lowered while maintaining its essentially straight form.

At the slope foot, a basal concavity evolves due to the slower rate of decline. This segment receives all the debris removed from the whole of the slope length, which accumulates on the footslope and protects the underlying bedrock from weathering. Transport of material through this zone is primarily by slopewash, and removal of debris is controlled by basal erosion. Where basal erosion is active, the concavity does not develop, but where erosion is limited, inputs to the footslope exceed outputs and the concavity is marked.

The Davisian view was that as the cycle progresses, slopes flatten. This was described as downwearing. A peneplain is a product not of lateral stream planation but of down wasting of interstream tracts, during the sequence of youth, maturity and old stage. The initial stage of valley development supports steep valley side slopes. Therefore, the rate of debris production is high, and the size of debris in transport over steep slopes is also very large. The transported debris accumulate at the break of slopes until the slope profile attains a balance between the rate of detritus production and its transport in the downslope direction. At an advanced stage of slope development, the slope angle and size of weathered debris both decrease but the thickness of waste mantle increases. In the old stage, all the slopes become gentle and the transport of debris which has become very fine, practically ceases. A thick cover of waste mantle covers the gentle slopes of the old stage. Thus, an initially steep straight slope is gradually replaced by a gentle convexo-concave slope profile in the old age stage.

Walther Penck provided an alternative hypothesis to the Davisian cyclic concept of slope development and lowering of relief. Penck began by assuming a straight, steep rock face (steilwand) standing above a stream at its base. Weathering of steilwand at a uniform rate throughout and removal of waste mantle from its foot cause it to retreat at a constant angle. The retreating steilwand initiates the development of upslope extending haldenhang or basal slope. The junction between steilwand and haldenhang is marked by a sharp slope discontinuity called piedmont angle.

Piedmont angle is maintained irrespective of the size of the backing scarp and the upslope extending basal slope surface. In due course of time, this basal slope itself becomes subject to weathering and retreats causing new slope gradient (geneigten flachen) to develop. Each geneigten flachen also undergoes a parallel retreat causing a slope segment of gentle gradient to develop at its foot when all the available relief is consumed by the process of parallel retreat of geneigten Lachen. This process leads to the formation of concave slope profile.

Theoretical Analysis

As a concept, the cycle of erosion that Davis envisaged, has not found wide acceptance among geomorphologists. The cycle concept is a part of history of geomorphology; it has not been used as the basis of geomorphological research since Davis' death. Rarely does one find the use of the term youthful, mature and old to describe landscape. There are many reasons for this—

1. Later studies basically concerned themselves with elaborating the mechanism of the process of erosion, where Davisian terminology finds no place.
2. Many regional studies have shown the complexity of the history of evolution of relief and have shown that relief usually has been formed under the influence of a series of partial cycles with in between climatic interventions.

Essentially, the cycle concept is a long term view of landscape, a view geological in it is scale. But, more than landscape development it stresses on:

1. Development of slope forms.
2. Profile and plans of rivers.
3. Development of drainage system.
4. Effect of rock type on different stages.
5. Importance of climatic accidents in controlling complexity of denudational forces.

One of the most attractive features of Davis cycle is its simplicity. This simplicity was achieved by eliminating supposedly irrelevant information and adopting restrictive assumptions. Thus, he made the landform system a closed system, which derived its initial input of energy in the form of potential energy from landform upliftment and operating against a background of static tectonics, base level, climate, hydrometeorology and vegetation. Such assumptions were basically made to facilitate his explanation of a more complex reality and it was wise of him to calculate that the very existence of his model was impossible without them.

While the importance of the cycle as a basis of research has often been exaggerated its influence on teaching in schools has been very great. In research, the related concept of denudation chronology and the recognition of erosion surfaces have been more widely used and still has much significance, especially in the Southern Hemisphere where the remnants of Gondwana land have many extensive erosional plains, or planation surfaces.

The Theory of Penck: Penck believed that landforms should be interpreted by means of ratios which might be expected to occur between erosional (exogenetic) processes and diastrophic (endogenetic) processes. Landscape was considered an expression of the phase and rate of upliftment in relation to the rate of degradation. Erosional processes were held to operate according to some worldwide laws whose rate were different in different climates.

These laws were:

1. Local intensity of erosion is directly related to the steepness of the slope segment.
2. The inclination of each segment of an erosional slope is determined by the sizes of the mobile debris.
3. The largest debris size which is mobile on a slope segment varies with the inclination of the slope. The greater the inclination, the greater is the largest debris size, which is mobile.
4. If the production of debris by weathering is uniform on a slope segment, erosion will, cause the segment to retreat parallel to itself.
5. If some eroded material is allowed to collect at the base of a retreating slope segment, a new segment with lower inclination will develop.

Penck was of the view that most tectonic movements began and ended slowly. The common pattern of such movements was a slow initial uplift, an accelerated uplift, a deceleration in uplift and finally quiescence. As there was regional updoming, there was a major phase of waxing development in which the accelerating uplift was more than the stream degradation and the ensuing landform was dominated by crustal instability.

Penck's view was that the characteristics of valley side slopes give a clue to the relative importance of exogenetic and endogenetic processes. The slopes of any region, according to Penck, are basically similar in gradient and their characteristics are determined by "intensity of erosion". Penck recognised three types of slopes—uniform or straight, concave and convex:

> Uniform slopes indicate constant intensity of erosion, i.e., uniform development.

Concave slopes indicate declining intensity of erosion, i.e., waning development.

Convex slopes indicate increasing intensity of erosion, i.e., waxing development.

The most important factor controlling intensity of erosion is the nature of crustal movement. Penck believed that slope gradients were largely determined by the intensity of erosion as influenced largely by crustal movement, but he also said that the details of slopes may reflect the character of rocks upon which they are cut.

Increasing Intensity of Erosion: This is the first stage of erosion. The initial surface (primarrumpf) is a featureless surface. As there is slow, long, continuous uplift, the primarrumpf retained its featureless character for a long time. As the rate of upliftment increased, stream incision led to obliteration of the featureless character and started developing narrow valleys. Thus, altitude and relief increased. Later on, the rate of upliftment increased tremendously which resulted in formation of steep V-shaped valley. Since the rate of upliftment was very high, downcutting or vertical incision could not keep pace with the accelerated uplift. Thus, the ridge summits and valley bottoms rose well above sea level. The slopes became convex because of extensive downcutting in valley bottoms without any corresponding denudation on the upper slopes.

Constant Intensity of Erosion: Constant intensity of erosion is divided into three phases:

Phase 1: There is an acceleration of uplift but the velocity of acceleration is not as high as compared to the previous phase. The summits rise but at a slower speed. The same is also true of valley bottoms. Valley deepening, valley bottom uplift and summit uplift are balanced in such a way that the relief remains constant. The valley sides are straight as both valley floors and summits are marked by matching erosion.

Phase 2: In this phase, the summits neither increase nor decrease, as the lowering of the summit height is offset by uplift which continues even in this stage. The valley deepening still continues at an accelerated rate because uplift provides the much needed stream velocity for vertical incision but this valley deepening is also matched by divide wasting, so the relief in this case, too, remains constant. But, the valley sides remain straight and retreat, parallel to themselves. As the rate of upliftment decreases, rate of valley degradation also decreases, and the valley floor rises. Summits too show the same tendency. The cumulative effect of these factors is that the relief remains constant and altitude is maximum.

Phase 3: At this point, there is a complete cessation of upliftment. The summit, thus, begins to lose height by erosion and there is a gradual decline of elevation. Since there is no upheaval to offset the effect of downcutting, valleys continue to become deep and their floors also become lower. But, uniform development is maintained as there is constant intensity of erosion and summit lowering is also as low. Hence, the relief remains constant. The whole stage is indicated by uniform slopes.

Declining Intensity of Erosion: By now, the upliftment of landform has completely stopped and erosion is dominant. Thus, declining landforms appear. The summits continue to be eroded and lowered, however, there is a considerable decrease in downcutting or vertical incision of valley floors. The net result is that altitude as well as relief both are lowered. At the same time, valley sides are retreating, parallel to themselves. The upper slopes maintain their angle of slope and valley floors increase in width. The surface from ridge crest to the stream (valley side slopes) now consists of two parts—

1. The upper part, called boschesteilwand or gravity slope, is relatively steep and maintains its angle by parallel retreat, although the height is declining.
2. The lower part, called haldenhang or wash slope, is gentle and has a declining gradient. The angle at which these slopes meet is distinct and it mainly depends upon the size of the material that is being moved over them and the amount of water available to move this material. The angle is fairly sharp in the form of a knick. The usually steep gravity slope or bosche retreats upslope maintaining its original angle. As it does so, the basal wash slope or haldenhang of lesser angle extended itself at the expense of the gravity slope.

The continuance of parallel retreat of gravity slope brings the crest of the ridges to lower altitude and, in particular, the steepest elements (boschungen), an angle of repose or free fall slope. The rapid parallel retreat of the boschungen soon consumes much of the convex, waxing slope and leaves behind at its base a slope element of lower inclination composed of talus material, susceptible to parallel retreat, under the action of creep and rainwash. The collection of mobilised finer material at the foot of the haldenhang may cause a lower slope segment to develop and so on.

The retreat of the boschungen gives rise, at a later time, to the production of steep sided residual inselbergs at the divides, and when these are consumed the whole landscape is made up of concave slopes of low angle and composed of slowly retreating slope segments. Such a surface is endrumpf.

The Criticisms

1. Penck made no clear distinction between continuous acceleration of uplift and continuous but intermittently accelerated uplift.
2. Valley side slopes do not mainly reflect the nature of crustal movement. They are undoubtedly affected by numerous other factors such as lithology, structure, climate, size of debris produced by weathering and its mode of transportation. Penck's cycle has become very complex on account of simultaneous upliftment and erosion.

Volcanism: Definition

Volcanism or volcancity includes all the phenomena associated with movement of molten material from the interior of the Earth to the surface. It is the process by which matter is transferred from the Earth's interior and erupted on to its surface.

In other words, it can be said to embrace all the activities and processes by which gaseous, liquid and solid substances of internal origin are ejected on the surface. Basically, there are three main processes—

1. Generation of magma,
2. Intrusion of masses of magma, and

3. Extrusion of molten material on to the surface by simple and quiet activity or by forceful ejection through explosive volcanic activity, breaking through the rocks.

The oldest rocks exposed on the surface of the Earth were produced by volcanic activity. So volcanoes, were active 3500 m.y. ago and, although, the intensity of activity has waxed and waned through the geological time, at present their products form most of the oceanic and part of the continental crust.

Volcanic activity is, undoubtedly, one of the greatest threats to life on Earth, and strange although it may sound, there would be no life on Earth without it, for the water brought from the Earth's interior by volcanic activity has allowed the creation of hydrosphere and atmosphere and has also been responsible for the evolutionary processes of life. The role of volcanism in crustal development requires an understanding of distribution of volcanoes in relation to major features of the Earth's surface—the ocean basins, mountain ranges, island arcs, continental margins and rift valleys. It is this relation, the geotectonic setting of volcanism, which determines both the formation and composition of magma, and hence, controls the amount of proportion of elements being added to the crust in a given region.

Belts of Volcanism

The most important volcanic belts can be grouped in five spatial classes.

Volcanic Ridges: The ridge volcanism, associated with worldwide system of ocean ridges, includes the Mid-Atlantic Ridge system, the East Pacific Rise, the Carlsbejg Ridge and the ridges encircling the Antarctica continent. Ridge volcanism takes the form of non-explosive fissure eruptions from active spreading ridge systems. This continuous and invisible kind of volcanism is responsible for almost all the crust beneath the oceans. The lavas are of basalt variety called tholeiite.

Volcanic Arcs: Arc volcanism comprises the island arcs and their associated volcanic activity, such as Aleutian islands, Kamchatka, Kurile islands, Japan, Philippines, Celebes, New Guinea, the Solomon islands, New Caledonia and New Zealand, the Indian Ocean girdle including Java, Bali and Sumatra, and Lesser Antilles, Scotia Tyrrhenian and Aegean Seas. These probably represent early stages of subduction zones. The active volcanoes tend to be rather evenly spaced at an average of 50-60 km, within more or less straight volcanic segments, varying from 100 km to 1100 km in length.

Volcanic Chains: Volcanic chains are straight line of volcanoes within the post-tectonic stage fold mountains and extend through the Andes of South America, Central America, Mexico and the Cascade mountains of the western USA The volcanic belts are strongly segmented, at least 7 segments are recognised between Guatemala and Costa-Rica.

Volcanic arc and volcanic chain

1. Circum Pacific
   - (i) West Pacific islands
   - (ii) West coast of North and South America
2. Indonesia
3. Mediterranean and SW Asia

Ocean Basins

1. Atlantic Ocean
2. Central Pacific
3. Indian Ocean

Continental interiors (East African Rift valley)

Apart from active volcanoes, the Pacific Ocean has some 10000 seamounts greater than 1 km in height and 100000 smaller abyssal hills, which tend to lie in lines.

Process of Volcanism

In all the regions having active volcanoes, a sequence of events takes place, i.e., generation of magma, its solidification below the surface and/or eruption through a central vent or volcanoes tend to occur singly within inactive belts with an average spacing of about 80 km. The kind of lava erupted from volcanic chains is principally of magma type intermediate between basalt and rhyolite and are called andesite.

Volcanic Clusters:

(i) These are found in oceans, for example, the Madeira, Galpagos, Canary, Azores in Atlantic, Reunion and Mauritius in Indian Ocean.

(ii) These are found on continents, for example, in Ethiopia, East African Rift valley, etc.

Volcanic Lines: Volcanic lines are lines of dominantly extinct volcanoes and seamounts linearly arranged. These are the Hawaiian-Emperor-Seamount chain, Line-Tuamotu chain and Austral-Marshall-Gilbert chain in the Pacific Basin.

There are more than 800 volcanoes, which are either active or dormant. These include:

80%

62%

45%

17 %

14%

4%

17%

13%

3%

1%

3%

through fissures leading to formation of various types of topographic features on the surface.

Magma: The term magma has been derived from a Greek word, which means kneaded mixture, i.e., Mike a paste'. In geological parlance, it refers to any hot mobile material within the Earth that is capable of penetrating into, or through the rocks of the crust.

Nature: Magma is a molten silicate material. It is not entirely liquid, but a combination of liquid, solid and gas. As it comprises many early formed crystals, it is slushy and it flows in a slow and sluggish manner.

The two most important constituents of magma are silica (SiO_2) and water (H_2O), and it is these two constituents, which control the properties of magma. Silica may constitute 37 to 75 per cent of magma. Other constituents are Oxygen (O), Silicon (Si), Aluminium (AI), Calcium (Ca), Sodium (Na), Potassium (K), Iron (Fe) and Magnesium (Mg). Dissolved gases, by and large, constitute a small percentage of magma. The most important of the dissolved gases, are volatiles, constituting 14 per cent of the total volume of the magma. These gases are important because by influencing the mobility and melting point, they determine the viscosity of magma and explosive characteristics of volcanic eruptions. The presence of volatiles increases the fluidity of magma and reduces its viscosity. Magma, which are rich in silica, are more viscous and offer resistance to flow.

Difference between Magma and Lava: Silicate melts (magmas) form in the lower crust and upper mantle, and that, under certain conditions, they can reach the surface to be erupted as lava. Lava differs from magma in containing fewer volatile components; and in cooling either in the air or under water, so that it crystallises rapidly. Magma which, having formed, remains deep within the crust, also cools, but only very slowly. In consequence, crystals are much better formed than in the rapidly cooled lavas.

Volatiles in Gases of Magmas

	Lava Gases from Mauna Loa and Kilauea Volcanoes (Per cent by Weight)	*Volatiles Free in Earth's Atmosphere and Hydrosphere (Per cent by Weight)*
Water, H_2O	60	93
Carbon, as CO_2 gas	24	5.1
Sulfur, S_2	13	0.13
Nitrogen, N_2	5.7	0.24
Argon, Ar	0.3	Trace
Chlorine, Cl_2	0.1	1.7
Hydrogen, H_2	0.04	0.07
Flourine, F_2		Trace

Magma can be divided into two classes— basaltic and granitic.

Basaltic magma contains about 50 per cent SiO_2 and has a higher temperature. Because of its lower silica content, basaltic magmas are more fluid. Although, basaltic magmas have a low silica percentage, they have high percentage of ferromagnesian minerals.

Granitic magmas contain 60-70 per cent SiO_2 and has temperature lower than 800°C. Because of higher silica content, they are thick and viscous. They contain felsic minerals. A third type of magma called alkaline lava has an unusually high ratio of sodium and/or potassium to silicon.

Variations in Properties Among Magma of Differing Compositions

Property	*Basaltic*	*Andesitic*	*Granitic*	
Silica content	Least (~50)	Intermediate (~60)	Most (~70)	
Viscosity	Least	Intermediate	Highest	
Tendency to form lavas	Highest	Intermediate	Least	
Tendency to form pyroclastics	Least	Intermediate	Highest	
Density	Highest	Intermediate	Lowest	
Melting point	Highesi	Intermediate	Lowest	
Typical minerals	Ca feldspar	Na feldspar	K feldspar	Pyroxene Olivine Amphibole Quartz Mica
		Pyroxene Mica	Amphibole	

Generation of Magma

Magma generation within the Earth is a result of complex interaction of increase in temperature, decrease in pressure and addition of water. Volcanic heat contributes about 1/100th of the total terrestrial heat loss, but it is very localised along belts of rising convective currents or above 122 odd hot spots, associated with rising plumes. The convecting layer probably occurs at depth of less than 700 km and contains a hierarchy of different cell sizes. About half of the heat in the convecting layer is generated within the layer itself and half comes from the mantle below. Magma rises within the Earth through piercing blobs or diapirs or as columns of partly molten rock by a complex set of processes.

The molten material that lies beneath the surface is called as magma and when the same molten material comes to the surface, it is called lava. Silicate melts (magmas) form in the lower crust and upper mantle, and that, under certain conditions, they can reach the surface to be erupted as lava. Lava differs from magma in containing fewer volatile components; and in cooling either in the air or under water, so that it crystallises rapidly. Magma which, having formed, remains deep within the crust, also cools, but only very slowly. In consequence, crystals are much better formed than in the rapidly cooled lavas.

At shallow depth, a combination of diapir pressure, thermal expansion and gas pressure (partly from groundwater turned to steam) may lead to surface eruption.

Magma generation in the upper mantle requires:

Increase in Heat: Melting in magma source areas requires thermal energy of about 100 cal/gm of magma produced. This energy may be supplied directly from:

1. Local concentration of radioactive elements and their decay, which emits a, (3 and y rays, thus producing heat.
2. Energy given off during Earthquakes.
3. Upward convection currents or plumes from the mantle.
4. Frictional heating in those places, where two surfaces move over one another.

Decrease Pressure: The temperature increases with depth at a rate of about 30°C/ 1000 km, establishing a temperature gradient which results in outward flow of heat from the Earth. However, the pressure on the overlying rocks also increases. Increase in pressure results in raising the melting point of all the rocks, so that at a level, the melting point of all rocks are raised above the prevailing temperature. At greater depths, the melting point would be so high that even allowing for heat during flow to the surface, the temperatures observed at the Earth's surface would have to be much higher than they, in fact, are. Since the melting point of silicates increases with load pressure, a decrease in pressure caused perhaps by faulting would reduce the effective melting point at a given depth resulting in melting.

Role of Water: Water increases the melting point of most silicates. When there is increase in water content (which may be due to breakdown of hydrous minerals carried into the subduction zone), the melting point is lowered, and hence, there is melting. However, the role of water is variable in the generation of magma.

With these concepts in mind, magma generation can be seen within the purview of plate tectonics, which accounts for the causation, distribution and type of volcanic activity.

Magma Generation and Plate Tectonic: The distribution of volcanic activity is not random. The location of active and recently active volcanoes points out that:

1. Volcanic activity coincides with seismic activity, which in turn, coincides with plate boundaries.
2. The type of volcanic activity depends on the type of plate boundary.

There are two types of associations of volcanic activity with plate tectonics – Plate margin volcanism and Intraplate volcanism.

Plate Boundaries of Volcanism Activity: There are two types of plate boundaries where volcanic activity takes place.

Constructive Margin or Divergent Plate Boundary: The asthenosphere, over which the lithosphere is floating, is made of peridotite. A slight degree of melting in the asthenosphere takes place, where temperature and pressure are just balanced for slight degree of melting to occur. It is because of this precarious temperature-pressure balance that the asthenosphere is soft and weak. As the plates break and move apart, i.e., diverge along a fissure or a fault, the asthenosphere also moves upward along the fault and the spreading zone. The disturbance of pressure-temperature equilibrium caused by decrease in pressure causes partial melting of peridotite to yield basaltic magma. The basaltic magma is extruded in a very quiet manner largely because-

1. Basalt has a low silica content and is very fluid, and
2. The eruption takes place all along the long fissure created by plate divergence.

This is how ridge volcanism takes place.

Destructive Margin or Convergent Plate Boundary: When two plates collide, the oceanic crust along with the marine sediments plunges into the mantle. As the higher density oceanic crust descends, it gets heated and melting starts. The partial melting of the basaltic oceanic crust alongwith oceanic sediments, rich in clay and silica, produces a magma that has a high silica content—this magma is called andesitic or granitic. Where subduction involves continental collision and mountain building, a greater variety of silica-rich rock develops from the partial melting of metamorphic rocks in the roots of the mountain systems.

The magma, thus generated, is of andesitic or granitic type and being lighter ascends to the surface through a vent. Since it has a higher silica content, the magma is viscous and flows out with explosive activity. Sometimes, the vent gets plugged by the silicic magma and, in the process of breaking it open, the accumulated force from within causes violent explosions. Thus, the volcanic activity at the convergent boundary is of 'explosive' type in contrast to the' quiet' type at the divergent plate boundary. Thus, where Nazca plate subducts under the South American plate, it sends volcanoes in Andes, the subduction of Pacific plate sends volcanoes in Bering, Kamchatka, Kurile, Japan, Philippines, New Zealand, the subduction of Indian plate similarly leads to volcanic activity in Java, Sumatra, and Bali, while Etna, Vesuvius and Stromboli were formed due to subduction of African Plate.

The segmentation of volcanic belts manifested by gaps, offsets or changes in volcanic products suggests that subducting plates may have been fractured into segments, some still actively ascending and capped by active volcanoes, the others stationary and capped by inactive ones. Active volcanoes tend to be rather evenly spaced on an average of 50-60 km.

The volcanic activity at plate margins, thus, can be easily explained either in terms of spreading activity or collision activity. However, volcanic activity at many places cannot be explained in these terms, such as Rift valley volcanism of East Africa, in the central Pacific in Hawaiian islands and Austral-Marshall-Gilbert islands. At all these places, volcanic activity does not take place at the margins, rather it take place within the plates.

Intraplate Volcanism: Volcanism in central parts of plates beyond constructive and destructive margins is not so common, but is the surface expression of local thermal variation or hot spots in the mantle. Intraplate volcanism occurs both in oceans as well as on continents.

Oceanic Intraplate Volcanism: Oceanic Intraplate volcanism includes Hawaii, Samoa, Galpagos and Easter Islands in the Pacific, the Azores, Madeira, Canary islands, St. Helena, Tristan da Cunha and Ascencin in the Atlantic, the Reunion and Mauritius in the Indian Ocean. Almost all these islands are members of larger group of islands or archipelago and volcanic activity is in pattern of clusters.

In the Pacific volcanoes have formed very long strings of islands and seamounts running for thousands of kilometres. The best examples are the Hawaiian-Emperor-Seamount chain, the Tuamotu-Line chain and the Austral-Marshall-Gilbert chain. These volcanic chains have been formed as the lithospheric plate moves over a mantle plume. Volcanism occurring over a hot spot produces a submarine volcano that grows into an island. If the mantle plume position remains fixed for a long time, the moving lithosphere carries the volcano beyond the magma source. This volcano then becomes dormant and a new one forms over the fixed mantle plume. A continuation of this process would build one volcano after another, producing a line or chain of volcanoes parallel to the direction of plate motion. This linear chain would encompass an active volcano at one end. Thus, Mauna Kialeua, at the extreme southern end of the Emperor Seamount chain, is an active volcano, so also is the Pitcairn island of Tuamotu group and Me Donald seamount of Austral Group. The lava erupted from intraplate oceanic volcanism is typically basaltic being of tholeiitic variety.

Continental Intraplate Volcanism: Continental Intraplate volcanism occurs along large intracontinental rifts. These are linear segments, in some cases exceeding 1000 km in length and generally about 30-60 km wide, along which the Earth's crust has subsided as much as 5 km along the crest of an elongated uplift. The East African rift system and Rhine graben are best known examples. The East African rift system has many volcanoes, Mt. Kilimanjaro, being the most prominent outside rift zones. Continental intraplate volcanism occurs in uplifted areas, such as Tibesti and Hogger in northern Africa, Massif Central in France, the Rheinish massif in

Germany, Yellowstone and Colorado Plateau in the USA and in eastern Australia.

Isolated volcanoes can result from small hot spots that do not last long enough to produce a volcanic chain. Alternatively, they can develop from minor pockets of magma carried with the moving asthenosphere.

Although, a majority of mantle plumes are associated with intraplate location, still quite a number of them lie on plate margins. Prominent mantle plumes that lie on mid-oceanic ridges or close to them are Iceland, the Azores and Tristan da Cunha. The volume of material that is ejected, exceeds the norm for mid-ocean ridges, and that is why all have emerged in the form of islands. The lava extruded is alkali-rich basalt, very typical of hot spots. Hot spots at converging plate boundary are difficult to locate. In these areas, volcanic activity is both abundant and complex, and it is difficult to isolate the contribution of hot spots from other sources of volcanism. Basalts, rich in alkali metals, are indeed found in some converging plate zones.

Violent Eruption of Some Volcanoes: Volcanoes may explode violently, or lava may flow out at the surface rather quietly. A single volcano may erupt explosively at one time and quietly at another; but in many volcanoes, one type of activity or the other predominates. The explosive activity of a volcano depends upon the amount of gas in the magma beneath it and the manner in which that gas escapes. An explosion cannot occur unless gas is abundant and has been trapped to build pressures to the bursting point. Much of this gas was formerly dissolved in the magma below the volcano. As magma cools, different minerals crystallise (are precipitated) from the molten solution, but dissolved gases tend to be excluded from such mineral grains.

Hence, the volume of liquid material in the magma body dwindles steadily as crystallisation proceeds, and the relative proportion of gas in the remaining liquid increases. When the gas content becomes too great, gas bubbles separate from the melt and stream upward in the magma chamber to zones of lower temperature and pressure near its roof. In a fluid magma—one that is very hot or that has a relatively low silica content—such gases may be able to escape readily and continuously without creating explosive pressures. As the gas expands, it forces the lava upward to the surface through cracks. If the magma is viscous, or if the volcanic vent has been plugged by solidification of once molten material, the gas cannot escape and pressure increase until an explosion occurs.

Some of the gases in the magma may come from the vaporisation of groundwater encountered by the magma in its upward journey or from rocks that have been melted and assimilated.

To illustrate, limestone (calcium carbonate) may add much carbon dioxide gas in this manner. Thus, the eruptive action of a volcano may change during its lifetime because the composition, temperature and amount of dissolved gas and magma are all variable factors.

Process of Volcanism

Once magma has been generated, it, being lighter than the adjoining rocks, moves upwards towards the surface. In the process it can:

1. Solidify below the surface
2. Break the surface open and start erupting on the surface.

In either case, it will give rise to certain topographic features. When the features have formed beneath the surface, they are called Intrusive, and when formed on the surface of the Earth, they are called extrusive.

Features Intrusive: The various intrusive features are either concordant or discordant with the bedding of the intruded sediments. Concordant means having boundaries parallel with bedding or foliation of the country rock, while discordant denotes margins that cut through the bedding or foliation of the country rock.

Major Features

1. Concordant: Lopoliths.
2. Discordant: Batholiths with associated bosses and stocks.

Minor Features

Laccoliths,

1. Concordant: Sills, Bysmaliths, Phacoliths.
2. Discordant: Dykes, Ring complexes to include cone sheets, ring dykes and cauldron subsidence.

Major Factors

Concordant

Lopoliths: Lopoliths (Greek: *'lopas'*-bowl, flat saucer or basin) are saucer like bodies concordant to the structure of the rocks into which they are intruded. They are of enormous size and are often layered, so that there is tendency towards the development of outward facing scarps with variations in the resistance of the various layers. The Bushveld lopolith in Transvaal, South Africa covers 55000 sq km and the Duluth Lopolith in Minnesota is some 40000 sq km.

Discordant

Batholiths: Batholiths or Bathyliths (Greek: *'bathos'*-depth) are very large deep-seated intrusions. In fact, they are the largest intrusive bodies of elongate oval or isometric shape that are usually found in central parts of folded regions. Since at the time of their formation, they lay deep within the crust, the magmatic material cooled very slowly with the result that these igneous rocks are coarsely crystalline. The mass of granitic rocks generally cuts across the surrounding rocks. In the simplest case, batholiths result from a single intrusion in which magma is injected into older rocks and cooled. Many intrusions may also be composed of several intrusions of similar composition.

Batholiths form exclusively on continents and large island arcs, and do not occur in oceanic islands. Batholiths are exposed on the surface only after considerable uplift and erosion, e.g., Idaho Batholith, which is exposed over an area of nearly 41000 sq km. Extensive exposure of batholiths are found in shield area of continents. These exposures are considered to be the roots of ancient mountain ranges, which have been eroded to lowlands. Therefore, the young mountain ranges and the ancient shields expose parts of batholith formed at different depth. Batholith landscape are characterised by such land form; as tors and inselbergs, although, their development depends very much on the nature of jointing.

Bosses and Stocks: Bosses and stocks are similar, but smaller intrusions, having an outcrop area (the area exposed at the surface) of less than 10 sq km. Compositionally, they are similar to batholiths. The main difference between bosses and stocks is that while bosses have circular intrusions, stocks have irregular intrusions. The granitic mass of Dartmoor is commonly called stock.

Sloping: Despite their huge sizes, batholiths do more upward. Even though, intruded rocks can be pushed upward by the slowly rising, there are other processes at work too. The rising magma apparently dislodges fragments of the overlying country rock by process known as sloping. Dislodged blocks are more dense than the rising magma and therefore, sink. As sinking proceeds, the fragments may react with and get partly dissolved.

Concordant

Xenoliths: Not all fragments dissolve, but instead they may sink all the way and reach the floor of the magma chamber. Any fragment of country rock, still enclosed in a triatomic body, when it solidifies, is known as xenoliths (Greek: *lxeno* '-stranger, *'lithos'-rock).*

Batholith Features

1. They are connected with central elevation of fold mountain region. Rarely are batholiths not associated with mountain ranges.
2. The form of batholiths is elongate and their long axis generally runs parallel to hundreds or more kilometres.
3. Batholiths have been intruded across the folds, indicating that they formed after the event of folding.
4. The depth of batholith formation is great—gravity measurements indicate that their downward extent is 10-15 km.
5. They are composed primarily of granite or granodiorite.

6. The upper surface of batholith is characterised by many domelike protrusions caused by stopping of magma.
7. Bathoiiths contain an enormous volume of rocks.
8. Batholiths have actually replaced the rock in which they have intruded, instead of having pushed them aside or upward. The molten material permeates the country rock (surrounding rock) at depth via existing voids. Sustained flow, magma buoyancy and anatexis or melting of country rock enlarge them into substantial plutons or underground reservoirs.
9. Large number of batholiths have discordant relationship, although some may be concordant.

Cupolas: Cupolas are small protrusions of subjacent batholiths exposed at the surface.

Concordant Type Factors

Sills: Sills are thin sheet-like intrusions of highly fluid basaltic magmas, injected between bedding planes, which separated layers of sedimentary rocks. They are commonly associated with less competent rocks, such as shales (i.e., rocks, which will deform easily under stress), hence, are concordant with the bedding planes. Sills range from a few centimetres to hundreds of metres thick and can extend laterally for several kilometres. Sills can be connected directly to a stock or batholith, or they can be local offshoots from dykes. Viscous magma rarely forms sills because viscous magma cannot be injected between older rocks without deforming them. Therefore, sills are always related to fluid basaltic magmas. One of the most extensive system of basic sills forms part of the Karroo series of South Africa, underlying 500000 sq km. Sills merge imperceptibly towards laccolith form, depending on fluidity of magma, the rate of cooling of magma at the extremeties of the sill, the resistance of rock to the injection of the magma and the pressure of the overlying mass of rock.

Laccoliths: Laccoliths (Greek: *'Lacos'-a* cistern), are formed when the viscous magma forces its way into overlying strata, which are bent upward to form a dome. Thus, laccoliths are characterised by a flat floor and an arched roof. Laccoliths have large range of sizes; they may be simple, compound or cedar treelike. Henry Mountains, Utah provides a classic example.

Bysmaliths: Bysmaliths are faulted variants of laccoliths, Newly injected magma sometimes increase the convexity of the laccolith until the rocks above are fractured along faults. Bysmaliths have considerable relief effects. These forms are only possible in areas of simple, near horizontal, sedimentary structures. Horizontal sediments of varying resistance give rise to plateau and escarpment ribbed by outcrops of more resistant beds.

Discordant Type Factors

Dykes: Dykes (Scottish-stone fence) are formed, when magma squeezes into fractures of the surrounding rock and cools vertically, cutting across the bedding planes because the emplacement of dykes is controlled by fracture systems within the surrounding rock. The width of a dyke can range from a fraction of a centimetre to hundreds of metres. The largest known dyke is the Great Dyke of Zimbabwe, which is 600 km long and has an average width of 10 km. In general, however, dykes are seldom more than 60 m wide. The fact that dykes are usually thin and often extend for considerable distance suggests that a fluid magma very rapidly invade fractures in the Earth's crust. The fractures may not have been in existence before the intrusion of the magma. It is more likely that the pressure exerted by the magma rapidly open fissures in an area already under tension. The tension, which may have created radial fractures, can be filled by magma to form radial dykes.. As magma forces its way upward, it sometimes pushes out a cylindrical section of the crust.

When magma solidifies as the circular or elliptical masses of rock that outline the cylindrical section of the crust, it forms ring dykes. The upward pressure of magma results in the formation of fractures in concentric sets. When these fractures get filled by magma, cone sheets are formed. Dykes in the form of radial pattern or parallel to each other around an igneous intrusion, are called dyke swarms and are found in Iceland and Greenland. Tertiary dyke swarms have caused the crust to stretch some 65 km (40 m) in the area of occurrence, such as in Iceland. Dykes that cause this type of stretching are termed dilation dykes.

Volcanic Necks: Volcanic necks are eroded remnant of solidified lava and/or agglomerate (cemented mixture of angular, fragmented material of volcanic origin) which formerly filled the vent of a volcano, but which has now been exposed by denudation of the surrounding cone. Shiprock, Mexico is one such neck.

Diapirs: Diapirs are developed from domes, when the roof rocks are ruptured and the intrusive body forces upwards.

Mafic: Intrusive features form, when the pressure of magma is not enough to break the surface open. The breaking of the surface is accomplished by enormous pressure from the rising magma. Through broken crust, many gases, lava (while inside the ground, the molten rock is called magma, when extruded on to the surface, it is called lava) and water come out. A violent explosion will also cause the blowing away of cone. The broken and fragmented portions are deposited along with lava to give rise to various surface features, collectively called extrusive features.

Volcanic Products: Physical products of volcanism includes volatiles, liquids and solids. Explosive volcanic eruption generates large volumes of gases and hot but solid rock fragments. Rapid gas expansion solidifies the rock fragments. The resulting mixture of hot, solid particles and gas, either may be blown upwards as huge ash clouds or avalanche rapidly flowing downslope as a fluidised flow. These clouds of glowing matter are termed Nuees Ardentees.

Volcanic products may be classified under three main headings: Volatiles, Lava and Solids.

Volatiles: Volatiles are elements and compounds, which are dissolved in a silicate melt and which would be gaseous at that temperature, had it not been for the elevated pressure and the solvent effect of the magma. The commonest volatiles are water and carbon dioxide.

Water: Some steam is normally produced when volcanoes are in their active state. If steam is emitted in huge quantity, its condensation produces heavy downpours. This rain, when it mixes with volcanic dust, results in mudflows known as lahars in Indonesia.

Gases: Apart from carbon dioxide, other materials include chlorine and hydrochloric acid; fluorine, hydrofluoric acid, many fluorides, including those of boron and iron; sulphur, sulphur dioxide; and in special circumstance hydrogen sulphide and exceptionally nitrogen, ammonia and borates may occur. The volatile constituents of magma are termed juvenile, if they are original constituents and resurgent, if they have been introduced as a result of contamination of the magma by material from country rocks.

Lava: Lava is the general name given to molten rocks coming out from volcanic vents or fissures. Lavas of three principal types are classified mainly by their proportions of silica.

Parent magma type	Class of lava	Kind of extrusive rock
Felsic	>70% SiO_2 (Rhyolitic)	Acidic Rhyolite
Intermediate	50-70% Intermediate (Andesitic)	SiO_2 Andesite
Mafic	<50% SiO_2 (Basaltic)	Basic Basalt

Although lavas have many similar components, no two volcanoes erupt lavas of exactly the same composition. In fact composition of lava m y vary from one eruption to another in the same volcano. Volcanoes, primarily erupting basaltic lavas, are common in ocean basins and in rift zones of the continents. Andesitic lavas erupt from volcanoes encircling the Pacific, where subduction zones occur. These lavas have been affected by their passage through the materials of the continental margins and have their original magmatic composition changed in the process. Siliceous (or, rhyolitic) lavas most commonly occur within continental regions, far from plate margins. In the Mediterranean Sea, a number of Italian volcanoes erupt lavas believed to have been contaminated by the ingestion of large quantities of limestone producing carbonatite lavas, hence, their compositions differ from those of their volcanoes. The world's only active carbonatite erupting volcano, O doinyo Lengal, occurs in East African Rift valley, in Tanzania. The lava is essentially liquid sodium carbonate that erupts black but soon weathers to white colour.

Basaltic lavas are of two types—An surface flow contains relatively little gas and is a slow moving, flow 3 to 10 m thick. The surface of the flow cools and forms a crust while the interior remains molten. This hardened upper crust is broken into jumbled mass of angular blocks and kinkers. The gas in the flow forms bubbles and vesicles and make the rock light and porous. Pahoehoe flow contains more gases, and hence, is more fluid. The higher fluidity makes it move faster. As the flow moves, it develops a thick glassy crust, which is moulded into pillow forms or can resemble coils of rope known as ropy lava.

Under some circumstances, especially when erupted under water, lavas consolidate as a pile of tightly packed rounded masses, a foot or so in diameter. These are known as pillow lava.

Solids: Fragments, blown out by explosive eruptions and subsequently deposited on the ground, are called Pyroclastic debris, which include (i) material thrown out of volcano as liquid globules which solidified in the air and are then deposited as solid particles, (ii) the solid material which has been derived from the fracturing and breakage of the volcano. They are erupted and deposited in two different ways; either as airfall deposits (tephra) or, as pyroclastic flows. The finest of the pyroclastic debris constitute dust (10-4 cm diameter). The falling debris are generally sorted with coarse and heavy particles depositing first and nearby, while the dust is winnowed away to fall last. The following tephra fragments settle around or near the volcano.

- *Ash:* Fragments consisting of sharply angular volcanic glass particles (shards) After consolidation it is known as tuff.
- *Cinders:* They are small slag like solidified pieces of lava.
- *Lapilli:* They are pieces about the size of walnut.
- *Blocks:* They are coarse angular pieces of the cone or masses, broken away from rocks that block the vent.
- *Bombs:* They are rounded masses or dots of lava that congeal while travelling through the air. They have twisted shape and spindle caused by aerodynamic modification.

- *Pumice:* They are solidified scum from the surface of the lava; it has a spongy or cellular character due to bubbles of steam and gas in the lava scum when it cooled. These are sufficiently buoyant to float on water.

Volcanic Products

Form	*Name*	*Characteristic (dimensions)*
Gas	Fume	
	Lavas	
Liquid	Aa	Rough, blocky surface
	Pahoehoe	Smooth to ropy surface
	Airfall fragments	
Solid	Dust	<1/16 mm
	Ash	1/16-2 mm
	Cinders	2-64 mm
	Blocks	>64 mm solid
	Bombs	>64 mm Plastic
	Pyroclastic flows	Hot fluidised flows
	Mudflows flows	Flows fluidised by rainfall, melting ice and snow.

Important Factors

Fluid Lava Spreads over Landscape: The products of eruption can be describes as the results of certain types of activity, such as exhalative (gas), effusive (lava), or explosive (tephra). Alternately, characteristic modes of activity and products can be defined by describing active volcanoes that behave in certain ways. The various extrusive features can be ascribed to any of the types of volcanic activity stated above.

Flood Basalts: Flood basalts form the simplest type of eruption. In fissure eruption, there is no explosive activity because the lava is basaltic and flows in a very fluid manner. The individual flows may be very extensive up to 100 km or more in length but comparatively only very much thicker. The step-like eroded edges of thick sections of superimposed flows led to basaltic rocks being termed as traps (German: *'trappean* '-steps). The most extensive features associated with basaltic extrusion are lava plains and plateaus resulting from the superimposition of vast series of individual flows of basaltic lava from systems of surface fissures. Such combinations of superimposed thin lava streams (individual flows being normally less than 10 m thick but very rarely reaching 100 m) can form extensive plains by drowning areas of low relief and disrupting drainage. Vast and persistent regional flows can produce lava plateaus thousand of metres thick.

Flood basalts represent the initial style of continental rifting and provide direct evidence of the nature of volcanic activity, along divergent plate margins. Where a spreading centre passes beneath a continent, the continental crust is split and large volume of basalt is extruded and spread over vast areas near the rift system. These great floods of lava fill lowlands and depressions in the existing topography and with subsequent uplift erode into basalt plateaus. Flood basalts are found in southern Brazil, Parana Plateau, Deccan Plateau of India, Columbia Plateau of the western USA and large areas of Siberia, Greenland, Antarctica and northern Ireland. Present day examples include Iceland flood basalt and Ethiopia associated with East African rift valley.

Fluid basaltic flow may enclose trees that are in the path of the flow. The wood burns, but in the process the lava congeals around the trunk, leaving a cylindrical-shaped tree mould where the tree formerly stood.

Sills and Lava Flows

Sill	*Lava Flows*
Uniform & Regular	Top generally irregular
Rarely vesicular or amygdaloidal	Commonly vesicular or amygdaloidal
Medium to fine grained, rarely glassy, owing to slower cooling	Fine-grained or glassy Owing to rapid cooling
Does not show weathering of upper surfaces	May show weathering of upper surfaces
Metamorphoses the upper and lower contact rocks	Can metamorphose the lower contact only
Stringers and veinlets of rock may penetrate both overlying rocks	Small veins in underlying rocks only. Care must be taken when flows have irregular tops
No fragments in adjoining sediments	Fragments may occur in overlying rocks, as pebbles or xenoliths

Shield Volcanoes: Shield volcanoes are huge domes of basalt formed as a result of continued outpouring of great quantities of highly fluid basaltic lava from a radiating series of fissures located on the oceanic floor or mid-oceanic ridge. The greatest assemblage of shield volcanoes is the Hawaiian islands, each of which is formed of one or more such volcanoes. The Hawaiian domes are the result of piling of basalts (alkaline olivine basalts) on the floor of pacific above a mantle plume. Measured upward from their bases on the ocean floor, these volcanoes are of the order of 8 km high. The lava dome has a lower slope of 2°- 4° and only 5°- 6° near the summit plateaus on which are located unrimmed subsidence craters, consisting of a sink or lava lake.

Columnar Joints: The interior of the basaltic flow may be massive and non-vesicular. As the flow cools, there is shrinkage in volume of thin sills and dykes producing long rock columns of prismatic form with four, five or six sides. This structure is called columnar jointing. For example, Giants Causeway, Ireland.

Pressure Ridges: In basaltic flow, the sides and top freeze, while the interior remains fluid. The pressure of gases from the fluid interior causes the crust of the flow to arch up in pressure ridges. They are typically 1m or so in height and 30 to 50 m long. Many pressure ridges have conspicuous cracks along the crests of ridges.

Lava blisters or squeeze ups are mounds formed by lava pushing up through the earlier formed crust.

Lava Tube: When the pressure is great enough, the fluid interior can break through the crust and flow out leaving a long lava tube.

Lava Tunnels: The surface of the lava flow cools quickly and solidifies but beneath this solid roof the basaltic lava is still very fluid and liquid in character. The solidified rocks now surround the liquid and act as an insulator keeping in the heat and preventing it from congealing. As the supply of lava stops, the liquid lava drains away leaving a long tunnel running down the middle of the flow.

Lavamites: Hot gases still remain in the lava tunnel after the last liquid lava has flowed away. The hottest are those closest to the roof. They may be so hot that they remelt the newly congealed lava. It then drips down producing formations like stalactites that have been called lavamites. The excess drips accumulate on the floor beneath forming slender upward growing pillars like stalagmites. Since the temperature within the tunnel is around 1000°C, no one has yet seen this happening.

Spatter Cones: After issuing from fissures, the basaltic lava usually spreads out over a large area. Along the fissures in some places, the rising lava may be concentrated and erupt like a lava fountain. The splashing of lava around the fountain can build up small conical-shaped mounds called spatter cones. Abundant on fissure eruption, spatter cones are sometimes also found on central type eruption.

Table Mountains: Table mountains are flat topped, volcanic cones with steep sides composed of pillow lava, palagonite and vitreous pyroclastic rocks. They form as a result of subglacial eruptions that melt the ice, creating a large pocket of steam and water beneath the ice. Eruption of lava into the water produces pillow lave, and interaction of the water and steam with the lava forms palagonite, which together built a flat topped cone. When the enclosing glacier melts, a table mountain occur in Iceland, where volcanic eruptions have taken place beneath glaciers.

Lava Dome, Cumulo Dome: When the lava, extruded from a central vent is very viscous (andesitic, rhyolitic), it does not flow but heaps up around the vent forming cylindrical and steep sided dome, which is more broader than higher. This is called cumulo dome. Since the vents are readily blocked, individual domes do not grow to great size and the effusive forms are modified by frequent explosions of trapped gases. The strange topographical forms produced by the clustering of small domes are very well seen in the Puys of the Auvergne, France.

Thoioid: Where the dome grows in pre-existing crater, it is known as thoioid. When there is successive flow of more acid lava, it generates steep sided mamelons.

Plug Dome: When a solidified lava plug is pushed bodily up out of the neck, like a piston, due to pressure of magma and volatiles trapped beneath it, Plug Domes up to more than 300 m high with no craters are created. Smaller plugs form lava spine, standing hundreds of feet above the summit of the dome. The two most famous spines are Puy de Dome Auvergne and Mount Pelee, Martinique.

Ash Plateau and Ignimbrite Plateau: When the flow consists of gas and suspended fragments of hot mineral grains, droplets of lava and pieces of rock, suspended like a dense dust cloud, are formed which move rapidly close to the surface. This type of eruption is called ash flow. Ash flows can reach velocities greater than 250 km/hr because the expanding gases constantly force the cooling lava particles apart. When the ash flow comes to rest, the particles of hot crystal fragments, glasses and ash fuse together to form welded tuff or Ignimbrite. These can be very large, for example, the Ignimbrite plateau of North Island of New Zealand and in Sumatra. As the cooling proceeds, the contracting mass develops columnar jointing.

Ash or Cinder Cone: Ash or Cinder cone is formed when the eruption is of central type with a predominance of pyroclastic materials. Growth of an ash or cinder cone begins around the vent with an encircling ring of pyroclastic materials consisting of ash, lapilli and coarser materials. This is called tuff ring, particularly, when composed largely of the finer sized materials.

Ash or cinder cones do not grow to a height of more than a few thousand feet. Such cinder cones are steep sided and rarely exceed 3000 m, as in Volcano de Feuge, Guatemala.

Composite or Strata Cone: Composite or strata cones exhibit rough stratification of pyroclastic and lava built up by alternate layers of lava and fragmental materials due to alternating explosive and quiet eruption. The sheets of solidified lava give strength to the edifice while the rock fragments that are ejected fall to build up a cone. Individual lava flows from the crater or flank of a cone form tongue like extensions down the cone called coulees. Most of the world's larger volcanoes, for example, Fujiyama in Japan, Vesuvius in Italy, Mt. Shasta in the USA, Cotopaxi in Ecuador, El Misti in Peru and Mayon in Philippines are examples of composite volcanic cone.

Parasitic Cone: Some cones are formed in the vicinity of the main cone and feed on the main cone. They are, thus, described as parasitic cones. Mt. Etna, Sicily has more than 200 parasitic cones.

Craters: A crater is a pit at the top of the volcanic vent. They are funnel-shaped depression, more or less circular in plan and rimmed by an infacing scarp. They may result either from explosive activity or from subsidence. Expulsion of volcanic ash, lapilli and other forms of ejecta may build a ring about a volcanic vent and produce a crater.

Roughly, circular explosion craters, occupied by lake, are commonly known as maars. They are quite widespread in the Effel district of Germany, New Zealand, Luzon and elsewhere. They are produced when surface water percolating downwards comes into contact with hot magma and is converted into steam. The resulting explosion blasts up to the surface blowing out a large hole in the ground in the process. Such craters are usually simple circular depressions surrounded by low rims of ejected debris and they may become occupied by lake.

Pit Craters: Pit crater or Volcanic sinks found on Hawaiian islands mainly result from collapse, following lowering of magma columns. They commonly consist of lava lakes.

Eruption Craters: Eruption craters are the holes at the top of the volcanic cone. When activity ceases, the crater becomes partially filled with debris.

Many volcanoes have a crater within craters. This is called nested crater, such as Mt. Taal, Vesuvius and Etna. These are formed by the diminution of volcanic forces after the first crater was formed.

Calderas: A caldera is a large depression, more or less circular in plan, with a diameter several times bigger than that of a crater. Calderas originate through collapse following eruption and the partial emptying of magma chamber. The rapid ejection of magma during a large pyroclastic eruption leaves the chamber from which the volcanic ash was emitted empty or partly empty. The now unsupported roof of the chamber slowly sinks under its own weight. Subsequent volcanic eruptions commonly occur along these fractures, thus, creating roughly circular rings of small cones. Calderas may also be formed by engulfment—i.e., the inward collapse of a volcanic cone, the result of molten lava being drawn off under the surface of the Earth or through a fissure in the flanks of a volcano. Crater Lake, Oregon, formed from a peak (now denoted by the name Mt. Mazama), partially destroyed by an eruption and accompanying engulfment 6500 years ago. It was once believed that this peak 'blew its top' by a paroxysmal eruption, but now thought that it collapsed inwards; of an estimated 130 cu km of material which, disappeared from the probable cone, only 14 cu km be accounted for as debris; the rest must have collapsed within the underlying magma.

They may be produced in three ways and are consequently of three types.

Krakatoa Type: This is formed by the explosion of the summit area.

Katmai Type: This is formed due to the magma drainage through adjacent conduits rather than through the main vent itself.

Valles Type: This is formed by the collapse, following the discharge of large volumes of ash and pumice flows from fissures, unrelated to pre-existing volcanoes.

Resurgent Couldrons: A volcano does not cease activity following the formation of a caldera. Magma starts entering the magma chamber, and in the process, causes the uplifting of the collapsed floor of a caldera to form a structural dome. Such a feature is called a resurgent couldron. Subsequently, small pyroclastic cones and lava flows build up in the interior of the caldera.

Lava Lake: Sometimes basaltic lava wells up inside a caldera that is so deep that the lava cannot escape. Then, it forms one of the most fearsome and dramatic of volcanic spectacles— a great lake of molten lava like the one that formed in the caldera of Nyiragango in Zaire.

The lava maintains a temperature of over 1000°C. Its red incandescence is seen on clear nights nearly 100 km away. Vast bubbles of gas, some over 13 km across, burst through the surface of lavá producing big waves.

Volcanoes do not remain active forever. Deep in the Earth's crust, there is some shift, the focus of intense heat moves slightly away and violent eruptions cease. But even though, the huge mass of magma may be cooling below, it is still capable of making its effect felt on the surface of the Earth.

Thermal Springs, Hot Springs or Warm Springs: Groundwater that has circulated to great depths in deeply folded rocks becomes heated. This heating may also be due to masses of magma that have pushed themselves very close to the crust, almost to the surface and are cooling. Groundwater coming in contact with this magma gets heated beyond 37°C, which when comes to the surface is known as Hot spring. These hot springs are found in many places in India—Ladakh, Manali, Sohna, Rajgir, Rajmahal, etc. Elsewhere, three volcanic regions have been known for their hot springs, i.e., Iceland, Yellow Stone National Park (USA) and North Island of New Zealand.

The hot acidic water from springs easily dissolves $CaCO_3$ from limestone. As the water cools the limestone deposits form travertines. Similar forms of siliceous deposits are called siliceous sinters.

Geyser: Geyser (Icelandic: *'Geyser '*-spouter or gusher) is a special type of hot spring which intermittently ejects steam and superheated water from an underground source through a hole in the ground.

The subterranean structure of a geyser comprises a number of water filled chambers, interconnected with a central pipe. Water, which accumulates in the pipe or vent, can warm up to such an extent that the water near the bottom of the vent begins to boil, causing the water higher up to blow into the air as a natural fountain. The erupted water falls back to Earth, trickles back into the vent and the warming process begins again, causing its intermittent flow. There are numerous geysers in Iceland and New Zealand. The best known example is the Old Faithful geyser in the USA, which erupts 'faithfully' every 66 minutes. Siliceous deposits formed around geysers are known as geyserite.

Fumaroles: Fumaroles are characterised by full scale continuous jet like emission of hot water from a vent. Such steam often carries sulphurous compounds and the sulphur condenses around the mouth of the fumaroles as brilliant yellow crystals called solftaras. Similarly, deposits of carbon dioxide are known as Moffetes and of boric acid as Saffoni.

Mud Volcanoes: Sometimes, the hot volcanic water is so acidic that it erodes the rock through which it passes breaking them down into tiny particles and carrying them up to the surface as mud. The colour of the mud depends upon the nature of minerals, which are dissolved. Sulphurous ones produce yellow, iron-rich minerals produce red, brown or black, and others create blue or grey. The various minerals combine in different proportions to produce such a range of colours that the pools of boiling mud are called volcanic paint pools.

Cycles of Geomorphology

The larger task of endogenetic forces is to create irregularities on the surface of the Earth by volcanism, mountain building, etc. As soon as these endforms are exposed on the surface, the various processes of weathering start working on them. Soon these are weathered, and in due course of time, the weathered products are transported by various agents. This process of erosion goes on and on, and the final result is the lowering of the elevation of the exposed surface nearest to the sea level or base level, (the level below which erosion cannot go on).

This obviously, takes time and the period of time taken is long. For this long a time, the land form is seldom stable to allow the erosion processes to go on. However, if the land form has been reduced to such a level, it usually happens that there is continued emergence of endogenetic forces and when the new landform has been created, the whole process is renewed again. The whole period, during which erosion processes erode the new surface to sea level is one cycle and since erosion plays an important part in it, it is called the Cycle of Erosion.

The cycle concept is basically an ideological framework because of the complexities in the intensities of endogenetic and exogenetic forces in time and space. Thus, the literal concept of cycle of erosion cannot be thought of. However, the significance of the cycle concept is that it explains the sequential development of landforms in a temporal framework.

There are three views on the cycle of erosion. These are of Davis, Penck and King.

These views relate to the sequential development of landforms in an orderly fashion during which the slope also evolves in a variety of ways. Thus, these cycles, while describing the development of landforms, also give information about the evolution of slopes.

Water from Ground

Water in greater or lesser amount is present everywhere in the soil, subsoil and bedrock: the part of the subsurface water that fully saturates the pore spaces is called groundwater. The water is of either external origin (derived from atmosphere or surface waters) or of internal origin (derived from interior of the Earth).

Significance of Flow

The rate of flow of groundwater increases as the slope of the water table increases, as long as the permeability of the ground remains uniform. Rate of groundwater flow through permeable materials is directly proportional to the product of the cross-sectional area through which flow can occur, the permeability, and the slope of the water table (the hydraulic gradient).

Climate's Role

Every Karst area has unique conditions of limestone lithology, relief and climate which give rise to unique assemblage of landforms and certain types can be identified.

True Karst or Holokarst: It occupies much of Croatia, Slovenia, parts of 'erstwhile Yugoslavia'. The limestones which have been involved in mountain building are thick and form ranges over 2000 m high. The amount of rainfall is very high 100-150 cm/yr. They are deeply fissured having numerous cavities.

Fluviokarst: Fluviokarst means where fluvial and karstic processes are working in combination. These are found in large areas of western and central Europe and the USA. Here, the limestones are less extensive and thinner than in Yugoslavia. Streams, thus, flow right across and water circulates less deeply. The soil covering is also thinner than in true karst region and this is due to increased slope wash. This area has also been affected by Quaternary changes.

Glaciokarst: Glaciokarst areas occur on the margins of or beneath ice sheets and many areas of very cold climate where snow accumulates and melts regularly. This leads to glacial scour with differential erosion followed by post-glacial solution, together with solution from glacier and snow melt streams.

Tropical Karst: Tropical Karsts are formed under conditions of high rainfall, and high temperature and include:

Kegelkarst or Cone Karst: It is a type of tropical karst terrain in which the characteristic landforms are numerous cone-like hills sometimes separated by cockpits (cockpit karst) which are enclosed depressions interconnected by sinuous corridors between the cones. Between the cockpits are smoothly crested steep sided hills often conical in shape. Kegelkarst are well developed in Indonesia, Jamaica, Puerto Rico, parts of Central America, Vietnam, south China and Malaysia. Turmkarst, also called Tower Karst or Pinnacle Karst, differs from kegelkarst largely because its genesis mainly depends on lateral solution caused by temporary or permanent water levees. It consists of steep sided hills or pinnacles called mogotes rising abruptly from an alluvial plain. It is well developed in Puerto Rico, central China, Vietnam, Sarawak, etc.

Phytokarst: A distinctive minor karst landform on Grand Canyon Island has been designated phytokarst. The surface of phytokarst is intricately pitted. Algae cover the surface and penetrate the karst surfaces from 0.1-1.2 mm depth.

Thermokarst: A term used by Russian geologists to designate Arctic coastal region and the associated features. However, they are not solutional in origin.

Polygonal Karst: It is found in New Guinea.

Development Features

Presence of soluble rocks, especially limestone, at the surface. Chalk and dolomite are also soluble but whereas dolomite is not as soluble, chalk lacks other conditions.

These soluble rocks need to be dense, highly jointed and thinly bedded. The highly jointed nature imparts permeability to the rock. If the rock is overall porous and permeable throughout, the whole of the rainfall will be absorbed. Permeability must be provided through joints and fissures. The absence of joints in the Chalk region is one of the reason why Chalk region of England and France have weakly developed karst features.

Existence of entrenched valleys below uplands which lie over soluble and well jointed rocks. Thus, ready downward movement of groundwater is facilitated. Good water circulation is of prime importance because moving water carries solution, standing water does not carry solution.

Rainfall must be at least moderate. Almost all the karst areas have moderate to abundant rainfall. Thus, arid and semi-arid regions do not display karst development. If karst regions exist at all in these arid and semi-arid regions, they must be relief features which have been formed when the climate was humid.

These favourable factors give rise to erosion and deposition which, in turn, gives rise to distinctive type of landform formation.

Surface: Fluvial Action

The features formed by erosion on the surface by fluvial action can be divided into two parts—

a. Surface solutional forms.

b. Closed depressions and individual hills.

Surface Solutional Forms

Terra Rossa: Where the slope is moderate to gentle, the descending solution of groundwater leaves a residue of red clayey soil on the surface and extending down into opened joints called terra rossa. The thickness of terra rossa varies from few metres to many and sometimes may cover the whole surface of the limestone region. It is very similar to laterite soils found in the tropics but it is not a variant of laterite and can be found not only in tropical and subtropical regions but as far north as southern Europe.

Lapies: In places of high relief, where the surface is not covered by terra rossa, i.e., when limestone is exposed at the surface, water running across the surface of blocks forms grooved, pitted, etched, fluted, rugged surfaces called lapies. Lapies are known by various names in different countries, e.g., karren in Germany and bogaz in various parts of former Yugoslavia.

Lapies are basically small forms which have been given a bewildering variety of names. There are various types of lapies or karren formation.

1. Rillenkarren are associated with steeper slopes under intense rainfall, usually without a tree cover. Here, the solution grooves are rather minute and closely spaced separated by sharp ridges. Rillenkarren are common in the Alpine karst and occur in larger forms upto 15-20 m long in the tropics.
2. Rund Karren are rounded grooves separated by rounded ridges formed due to the passage of water between soil and rock.
3. Kluftkarren are related closely to jointing. Grikes (deep clefts or fissures) appear as cleft-like ruts.
4. Trittkarren are smooth surfaces with small steps and solution basins (kammenitsa).
5. Spitskarren are formed under conditions of rapid transient flow on shedding surfaces where depths are limited and the soluble elements are dissolved giving rise to dissected surface called spits karren.
6. Deckenkarren are shallow forms formed by direct action of plants.
7. Hohlkarren are formed under peat conditions.
8. Rinnenkarren have sharp crests and inclined surfaces.

Limestone Pavement: A Limestone pavement is a glacially planed and smoothed surface of bare limestone (glaciokarst), which has subsequently been dissected by solution resulting in the enlargement of vertical joints (grikes) to produce flat topped rock—clints.

Different Types of Drainage

Subterranean Cut-off: A subterranean cut-off is an underground diversion of a part or the whole of a surface stream beneath a meander spur along an entrenched valley through natural tunnels.

Natural Tunnels and Bridges: Natural tunnels make subsurface connection between surface fluvial features either by circuiting a meander or expanding its dimension. These may be formed either as a result of subterranean cut-off development or due to stream piracy.

When a large section of cave roof falls and a middle section is left standing, a natural bridge may be formed. Natural bridges may also be formed when surface river disappears into a fissure in the bedrock which runs underground a short distance, and then gushes out on the face of a cliff. As the fissure is enlarged, the area between it and the cliff is left as a natural bridge. The Natural Bridge of Virginia is believed to have formed in this manner.

Hums: Hums are residual hills analogous to monadnocks of normal river erosing which remain after the karst landscape has fully developed. These hills are called haystack hills or Pepino hills in Puerto Rico and mogotes in Cuba.

Sources of Groundwater

There are four sources of groundwater:

Connate Water: Water entrapped in the interstices of a sedimentary rock at the time the rock was deposited.

Meteoric Water: Meteoric water originates in the atmosphere, falls as precipitation and becomes groundwater by infiltration. Meteoric water constitutes the bulk of groundwater and this is evident in the fluctuation of water level in wells during the rainy season when the level goes up, and in summer season when the level goes down.

Juvenile Water: Juvenile water is that water which is considered to have been generated in the interior of the Earth and have reached the upper levels of the Earth's surface for the first time. It is also called magmatic water.

Condensational Water: Condensational water is the basic source of replenishment in deserts and semi-desert areas. In the deserts, particularly in summer, the land is always warmer than the air in the soil. This results in a difference of pressure between the water vapour in the atmosphere and the water vapour in the soil. The water vapour from the atmosphere penetrates into the rocks and is converted into water as the temperature of the water vapour drops below. Thus, certain amount of water may accumulate in rocks in desert and semi-desert region. All the above sources may get mixed along complex migration routes.

Depth of Groundwater

Water is present everywhere beneath the ground surface. More than half of all groundwater, including most of the water that is usable, occurs within about 750 m of the Earth's surface. The volume of water in this zone is estimated to be equivalent to a layer of water approximately 55 m thick spread over the world's land area. Below a depth of about 750 m, gradually though irregularly, water decrease in amount. Holes drilled for oil have found water as deep as 9.4 km. A deep hole drilled by Soviet scientists on the Kola Peninsula, encountered water at depths of more than 11 km. However, even though water may be present in crustal rocks at such depths, the pressure exerted by overlying rocks is so high and openings in rocks are so small that it is unlikely that much water is present or that it can move freely through the enclosing rocks.

Occurrence of Groundwater

Underground water, groundwater, subsurface water and subterranean water are all general terms used to refer to water in the pore spaces, cracks, tubes, and crevices of the consolidated and unconsolidated materials beneath the surface. Water below the ground surface occurs in four zones-soil zone, intermediate zone, capillary zone and saturation zone. Water that moves down from the surface is caught by rock and Earth materials and is checked in its downward progress. The zone in which this water is held is known as the zone of aeration. The spaces filling particles in the zone are filled partly by water and partly by air. Two forces operate to prevent suspended water from moving deeper into the

Earth (i) the molecular attraction exerted on the water by the rock and Earth materials, and (ii) the attraction exerted by the water particles on one another.

The zone of aeration is subdivided into three zones - soil moisture zone, intermediate zone and capillary zone. The soil, intermediate and capillary zone are called Vadose Zone (Latin: 'vadosus '-shallow). Some of the water that enters the zone of soil moisture, from the surface, is used by plants and some of it is evaporated back into the atmosphere. Some water passes down into the intermediate zone where it is held by molecular attraction (suspended water). Little movement occurs in the intermediate belt. The movement of water here takes place under the influence of gravity. Capillary this zone may be absent or may be several hundred metres thick where the intermediate belt is absent. The soil moisture zone is directly above the capillary zone.

At the base of the intermediate zone, is the capillary fringe, a thin layer (ranging from few cm to 2 or 3 m) in which water has been drawn upwards by capillary force.

The capillary state of water is temporarily destroyed when heavy rain or rapidly melting snow allow large amounts of water to infiltrate. At such times, the soil openings are fully saturated and the water moves downwards under the influence of gravity to reach the water table. In this way, the groundwater body is replenished by a process called recharge.

Beneath the zone of aeration, lies the zone of saturation. This is also known as the phreatic zone. Here, the openings in the rocks arid Earth materials are fully saturated with water. This water constitutes the groundwater and the surface between the zone of aeration and zone of saturation is called the groundwater table or simply water table. The level of the groundwater table fluctuates with variations in the supply of water coming down from the zone of aeration. The depth of the water table at a given place can be determined by noting the level at which water stands in a well of large diameter. The maximum water elevation in a well penetrating the groundwater zone is called the piezometric water table. The thickness of the zone of aeration differs from one place to another and the level of water table fluctuates accordingly. In general, the water table tends to follow the irregularities of the ground surface reaching its highest elevation beneath hills and its lowest elevation beneath valleys.

A geological formation saturated by water that will yield appreciable quantities of water that can be economically used and developed, is called an aquifer (Latin; to bear water), for example, a sandstone layer, when high in porosity and permeability, can hold and transmit large quantities of water. Unconsolidated sand and gravel, sandstone, limestone, lava flows, and fractured plutonic and metamorphic crystalline rocks are examples of typical aquifers.

An aquifer can be either confined or unconfined. An unconfined aquifer extends continuously from the land surface downward through material of high permeability. Recharge to the aquifer may be from percolation downward through the unsaturated cone, from lateral groundwater flow, or from upward seepage through the underlying material. In other words, water table is free to receive recharge from above and can rise or fall freely within the aquifer in response to changes in the amount of recharge received and where upper boundary of the saturation zone is the same as water table. Confined aquifers, also known as artesian or pressure aquifers, are overlain by confining layers having an impermeable geological stratum that keeps water under pressure. Hydrostatic pressure in this confined aquifer is sufficient to raise water in wells to levels higher than the upper surface of aquifer. This upward water movement is called artesian flow.

Confining layers can be subdivided into aquitards, aquicludes and aquifuges. Aquitards consist of confining bed that retards but does not completely stop the flow of water to or from an adjacent aquifer. Although, it does not readily release water to wells or springs, it may function as a storage chamber for groundwater. An aquiclude (Latin: water and close or shut) is a formation that stores water but is incapable of transmitting it. Clay, shale and most metamorphic and igneous rocks are poor aquifers and water passes through them very slowly, if at all. Aquicludes block the downward percolation of water in the hill and lead to the building up of small saturated zone above the impermeable layer. Because the local water table here is actually above the main water table, it is called a perched water table. The water that flows laterally along this, impermeable rock may emerge at the surface as spring.

A rock body or rock layer having no interconnected openings or interstices and which, thus, that neither transmits nor stores water is an aquifuge, for example, obsidian and quartzite.

Water enters an aquifer through a recharge area, which is an area where the water bearing stratum is exposed to the atmosphere or is overlain by a permeable zone of aeration. A water well can withdraw groundwater by digging or drilling through the zone of aeration into the zone of saturation where water flows from pore spaces into the well. Without pumping the well, water will rise to the level of the water table. In order to extract the water, pumping is usually needed to bring it to the surface. However, under pumping, the removal of water cause the water table to be drawn down around the well, producing a cone of depression in the water table. If the rate of withdrawal of water by pumping is greater than the inflow of groundwater to the well, the zone of depression will grow increasingly large, flattening the hydraulic gradient in the vicinity of the well and progressively decreasing the rate at which water can be withdrawn, until ultimately the well goes dry.

Conditioning Factors for the Occurrence of Groundwater: The occurrence of groundwater is controlled by:

Climate: Groundwater, generally occurs at great depths in arid regions and at shallower depths in humid regions. In hilly regions, groundwater does not occur at such a shallow depth while in valleys, it is at or near the surface. In rainy season, the water table rises while in dry season it sinks.

Topography: The water table is higher under highest areas of surface-hilltops and divides-but descends towards the valleys where it appears at the surface close to streams, lakes or marshes. This is because water percolating down through the unsaturated zone tends to raise the water table, whereas seepage into streams, swamps and lakes tend to draw off groundwater and lower its level.

Bibliography

Arvill, R.: *Man and Environment Crisis and the Strategy of Choice,* Penguin, New York, 1967.

Ashok, K.: *Cities of the World the, World Regional Urban Development,* Harper Collins, New York, 1993.

Bambrick, S.: *The Cambridge Encyclopedia of Physical Geography,* Cambridge University Press, New York, 1994.

Bennett, D. G.: *World Population Problems,* Park Press, Campaign, 1984.

Bhardwaj, N.: *Elements of Ecology,* Vikas Publications, New Delhi, 1979.

Bhatia, B. M.: *Poverty, Agriculture and Economic Growth,* Vikas Publishing House, New Delhi, 1977.

Billihgs, W. D.: *Plant, M[illegible]n and the Ecosystem,* MacMillan, Lond[illegible] 1971.

Biswas, M. R.: *Desertification: Environmental Science and Application,* Pergamon Press, New York, 1980.

Bonar, James: *Malthus and His Work,* [illegible]llen and Un[illegible], London, 1885.

Brun, S. D[illegible]s *of the World: World Regional Urban Development,* Harp[illegible]Row, New York, 1983.

Carlson, L.: *[illegible]ography and World Politics,* Prentice Hall, Englewood, 1958

Clark, W.: *Migration and Development,* The Hague, Mouton, 1975.

Clarke, C. D.: *Geography and Ethnic Pluralism,* Allen and Unwin, Londo[illegible], 1984.

Clarke, J. I.: *Geography and Population: Approaches and Applications*, Pergamon Press, New York, 1984.

Cole, J.: *Development and Underdevelopment: A Profile of the Third World*, Methuen, London, 1987.

Cressey, G. B.: *Asia's Land and Peoples*, McGraw-Hill, New York, 1963.

Critchfield, H. J.: *General Climatology*, Prentice Hall of India Private, New Delhi, 1979.

Demko, G. J.: *Population Geography: A Reader*, McGraw-Hill, New York.

Doshi, J. K.: *Bhils: Between Societal Self Awareness and Cultural Synthesis*, Sterling, New Delhi, 1971.

Doxiadis, C. A.: *Ekistics, An Introduction to the Science of Human Settlements*, Oxford University Press, New York, 1968.

Febure, L.: *Geographical Introduction to History*, MacMillan, London, 1925.

Frankel, F. R.: *India's Green Revolution: Political Costs of Economic Growth*, Princeton University Press, Princeton, 1971.

Geoffrey J.: *All Possible Worlds: A History of Geographical Ideas*, John Wiley, New York, 1993.

Gottman, J.: *A Geography of Europe*, Holt, Reinehart and Winston, New York, 1969.

Goudie, A.: *The Human Impact on the Natural Environment*, Basil Blackwell, Oxford, 1986.

Gregor, Howard F.: *Geography of Agriculture: Themes in Research*, Prentice Hall, New Jersey, 1970.

Gregory, D.: *Ideology, Science and Human Geography*, Hutchinson, London, 1978.

Haddon, A. C.: *The Races of Man and their Distribution*, Oxford, New York, 1925.

Hall, D.: *A History of South East Asia*, St. Martin, New York, 1968.

Haq, M. U.: *Human Development in Changing World*, Oxford, New York, 1992,

Hartshorn, T.: *Interpreting the City: An Urban Geography*, John Wiley and Sons, New York, 1990.

Hawkes, J. J. and Woolley, L.: *Prehistory and the Beginnings of Civilization*, Allen and Unwin, London, 1963.

Heathcote, R. L.: *The Arid Lands: Their Use and Abuse*, Longman, New York, 1983.

Hiro, Dilip: *Dictionary of the Middle East*, St. Martin's Press, New York, 1996.

Howe, G. M.: *The Soviet Union: A Geographical Study*, Longman, New York, 1986.

Husain, M.: *Agricultural Geography*, Inter India Publications, New Delhi, 1979.

Isard, W.: *Methods of Regional Analysis-An Introduction to Regional Science*, MIT Press, Massachusetts, 1960.

Jackas, G.V. and White, R.O.: *The Rape of the Earth*, Faber, London, 1939.

Johnson, B.: *India: Resources and Development*, Arnold -Heinemann, London, 1980.

——————: *South Asia*, Heinemann Educational Books Ltd., London, 1969.

Johnston, R. J.: *A Dictionary of Human Geography*, Black-well, Oxford, 1990.

King, Russel: *The Mediterranean Environment: Environment and Society*, John Wiley, New York, 1990.

Kosinski, L. A.: *People on the Move*, Methuen, London. 1975.

Kothari C. R.: *Research Methodology, Methods and Techniques*, Wiley Eastern, New Delhi, 1990.

Lal, D. S.: *Climatology*, Chaitanya Publishing House, Allahabad, 1986.

Leeming, F.: *The Changing Geography of China*, Blackwell, Cambridge, 1993.

Ley, D.: *Humanistic Geography: Prospects and Problems*, Croom Helm, London, 1978.

Livingstone, D.: *The Geographical Tradition,* Blackwell, Cambridge, 1992.

Major, S.: *The Land and People of China,* Lippincott, Philadelphia, 1989.

Majumdar, D. N.: *Races and Cultures of India,* Asia Publishing House, Bombay, 1958.

Manmion, A.M.: *Global Environmental Change,* Longman, New York, 1991.

Martin, G. J.: *All Possible Worlds: A History of Geographical Ideas,* John Wiley and Sons, New York, 1993.

Mason, D.: *Theories of Races and Ethnic Relations,* University Press, Cambridge, 1986.

Mcneill, William H.: *Rise of the West: A History of Human Community,* University of Chicago Press, Chicago, 1970.

Mehta, J. K.: *Present State of Indian Economy,* Oriental Publishers, Allahabad, 1977.

Moore, A.: *Advances in World Archaeology,* Oxford, New York, 1985.

Mukherji, Shekhar: *Dynamics of Population and Family Welfare,* Himalaya Publishing, Mumbai, 1991.

Mumford, L.: *The City in History,* Harcourt, Brace and World, New York, 1961.

Muthiah, S.: *An Atlas of India,* Oxford University Press, New York, 1992.

Myrdal, Gunnar: *Rich Lands and Poor,* Harper and Brothers, New York, 1957.

Ojha, S.: *Geography of India,* Bhaugolik Adhyayan Sansthan, Allahabad, 2001.

Pannell, C. W.: *East Asia,* Rowman and Littlefield, Lanham, 1999.

Pannikar, K. M.: *A Survey of Indian History,* Popular Prakashan, Bombay, 1964.

Park, C.: *Ecology and Environmental Management,* Butterworths, London, 1980.

Parmer, B. H.: *The Green Revolution in India,* The MacMillan, London, 1980.

Persson, G. A.: *Resources and World Development,* John Wiley, Chichester, 1987.

Pfajjlin, J. R. and Ziegler, E. N.: *Encyclopedia of Environmental Science and Engineering,* Gordon and Breach Science Publishers, New York, 1976.

Pilgrim, G. E.: *Agro Climatic Regional Planning: An Overview,* Government of India, New Delhi, 1989.

Pohekar, G. S.: *Studies in Green Revolution,* United Asia Publications, Bombay, 1970.

Potter, R. B.: *Geography of Development,* Longman, New York, 1999.

Qureshi, M. H.: *India: Resources and Regional Development,* N. C. E. R. T., New Delhi, 1990.

Rampino, M. R.: *Climate: History, Periodicity and Predictability,* Von Nostrand Reinholt, New York, 1987.

Rao, K. L.: *India's Water Wealth,* Orient Longman, New Delhi, 1975.

Rapp, A.: *Human Activity and Environmental Processes,* John Wiley, Chichester, 1987.

Raza, M.: *Development and Ecology,* Rawat Publications, Jaipur, 1992.

Robert, W. K. and Gilbert, F. W.: *The Environment as Hazard,* Oxford University Press, New York, 1978.

Rostow, Walter W.: *The Stages of Economic Growth,* Cambridge University Press, London, 1971.

Rubenstein, J.M.: *The Cultural Landscape,* Prentice Hall, New Delhi, 1990.

Satya, Rai M.: *Partition of the Punjab,* MacMillan, London, 1965.

Sauer, C.O.: *Agriculture Origins and Dispersals,* American Geographical Society, New York, 1952.

Sealey, Neil: *Caribbean World: A Complete Geography,* Cambridge University Press, New York, 1992.

Sen, Bandhudas: *The Green Revolution in India: A Perspective,* Wiley Eastern, New Delhi, 1974.

Sharma, P. D.: *Elements of Ecology,* Rastogi, Meerut, 1988.

Shaw, d. J.B.: *The Post Soviet Republics: A Systematic Geography,* John Wiley and Sons, New York, 1992.

Singh, Abha Laxmi: *The Problem of Wastelands in India,* B. R. Publishing Corporation, Delhi, 1985.

Smart, Ninian: *The World's Religions,* Cambridge University Press, New York, 1998.

Smith, C. J.: *China: People and Places in the Land of China,* Sharpe, New York, 1984.

Sopher, D. E.: *The Geography of Religions,* Prentice Hall, Englewood Cliffs, 1967.

Spate, O.H.K.: *The Spanish Lake: A History of the Pacific Since Magellen,* Beckenham U.K., Croom Helm, 1979.

Srinivasan. T.N.: *The Green Revolution,* Discussion Paper 66, Calcutta: Indian Statistical Institute, 1971.

Stamp, L. D.: *Asia: Regional and Economic Geography,* MacMillan, London, 1958.

Stock, R.: *Africa South of the Sahara: A Geographical Interpretation,* Guilford Press, New York, 1995.

Strahler, A. N. and Strahler, A.H.: *Grography and Man's Environment,* John Wiley, New York, 1976

Streeten, P.: *Global Governance for Human Development,* Oxford, New York, 1992.

Sukhwal, B. L.: *India: Economic Resource Base and Contemporary Political Patterns,* Sterling Publishers Private Limited, New Delhi, 1987.

Sundaram, K. V.: *Geography of Under-development: The Spatial Dynamics of Under-development.* Concept, New Delhi, 1983.

Taylor, G.: *Geography in the 20th Century,* Methuen, London, 1967.

Taylor, James A.: *Weather and Agriculture,* Pergamon., Oxford, 1967.

Thomas, J.: *Urban Geography: A First Approach,* Wiley and Sons, New York, 1982.

Thomson, R. D.: *Processes of Physical Geography,* Longman, London, 1986.

Trewartha, G. T.: *An Introduction to Climate,* McGraw-Hill Book Company, New York, 1954.

Trewartha, Glenn: *A Geography of Population: World Patterns,* John Wiley and Sons, New York, 1969.

Turk, J.: *Introduction to Environmental Studies,* C.B.S. College Publishers, Chicago, 1985.

Unwin, Tim,: *A European Geography,* Harlow, New York, 1998.

Van Eckelen, W. F.: *Indian Foreign Policy and the Border Dispute,* The Hague, London, 1967.

Wadia, D. N.: *Geology of India,* Tata McGraw-Hill Publishing Co., New Delhi, 1975.

Wadia, Mehar D. N.: *Minerals of India,* National Book Trust, New Delhi, 1994

Ward, Paul: *World Regional Geography: A Question of Place,* Harper and Row, New York, 1977.

Wegener, A.: *The Origin of Continents and Oceans,* John Biram, New York, 1966.

Wheeler, James. O. and Muller, Peter O.: *Physical Geography,* Wiley, New York, 1981.

White, Paul: *The West European City: A Social Geography,* Longman, London, 1984.

Wurfel, D. and B. Burton,: *South East Asia in the New World Order,* St. Martin's Press, New York, 1990.

Zelinsky, W.: *A Prologue to Population Geography,* Prentice Hall, New Jersey, 1966.

Zelinsky, W.: *Geography and a Crowding World,* Oxford University Press, New York, 1970.

Zeuner, F.E.: *A History of Domesticated Animals,* Hutchinson, London, 1963.

Zimmermann, E.W.: *World Resources and Industries,* Harper and Row, London, 1951.

Zohry, D. and M. Hopf: *Domestication of Plants in the Old World,* Clarendon Press, Oxford, 1988.

Index

B

C

D

E

H

I

J

L

M

N

O

P

R

S

T

U

V

W

X

Y

Z

□□□